I0066234

Fiber Optic Sensing - Principle, Measurement and Applications

Edited by Shien-Kuei Liaw

Published in London, United Kingdom

IntechOpen

Supporting open minds since 2005

Fiber Optic Sensing - Principle, Measurement and Applications
http://dx.doi.org/10.5772/intechopen.78479
Edited by Shien-Kuei Liaw

Contributors
Fathy Mohamed Mustafa, Moferh Toba, Avishay Shamir, Amiel Ishaaya, Aviran Halstuch, Shouhei Koyama,
Hiroaki Ishizawa, Cheng Feng, Thomas Schneider, Jaffar Kadum, Shien-Kuei Liaw

© The Editor(s) and the Author(s) 2019
The rights of the editor(s) and the author(s) have been asserted in accordance with the Copyright,
Designs and Patents Act 1988. All rights to the book as a whole are reserved by INTECHOPEN LIMITED.
The book as a whole (compilation) cannot be reproduced, distributed or used for commercial or
non-commercial purposes without INTECHOPEN LIMITED's written permission. Enquiries concerning
the use of the book should be directed to INTECHOPEN LIMITED rights and permissions department
(permissions@intechopen.com).
Violations are liable to prosecution under the governing Copyright Law.

[(cc) BY]

Individual chapters of this publication are distributed under the terms of the Creative Commons
Attribution 3.0 Unported License which permits commercial use, distribution and reproduction of
the individual chapters, provided the original author(s) and source publication are appropriately
acknowledged. If so indicated, certain images may not be included under the Creative Commons
license. In such cases users will need to obtain permission from the license holder to reproduce
the material. More details and guidelines concerning content reuse and adaptation can be found at
http://www.intechopen.com/copyright-policy.html.

Notice
Statements and opinions expressed in the chapters are these of the individual contributors and not
necessarily those of the editors or publisher. No responsibility is accepted for the accuracy of
information contained in the published chapters. The publisher assumes no responsibility for any
damage or injury to persons or property arising out of the use of any materials, instructions, methods
or ideas contained in the book.

First published in London, United Kingdom, 2019 by IntechOpen
IntechOpen is the global imprint of INTECHOPEN LIMITED, registered in England and Wales,
registration number: 11086078, The Shard, 25th floor, 32 London Bridge Street
London, SE19SG – United Kingdom
Printed in Croatia

British Library Cataloguing-in-Publication Data
A catalogue record for this book is available from the British Library

Additional hard and PDF copies can be obtained from orders@intechopen.com

Fiber Optic Sensing - Principle, Measurement and Applications
Edited by Shien-Kuei Liaw
p. cm.
Print ISBN 978-1-78984-625-6
Online ISBN 978-1-83962-181-9
eBook (PDF) ISBN 978-1-83962-182-6

We are IntechOpen,
the world's leading publisher of
Open Access books
Built by scientists, for scientists

4,300+
Open access books available

116,000+
International authors and editors

125M+
Downloads

151
Countries delivered to

Our authors are among the

Top 1%
most cited scientists

12.2%
Contributors from top 500 universities

CLARIVATE ANALYTICS
BOOK
CITATION
INDEX
INDEXED

WEB OF SCIENCE™

Selection of our books indexed in the Book Citation Index
in Web of Science™ Core Collection (BKCI)

Interested in publishing with us?
Contact book.department@intechopen.com

Numbers displayed above are based on latest data collected.
For more information visit www.intechopen.com

Meet the editor

Professor Shien-Kuei Liaw received a PhD in photonics engineering from National Chiao-Tung University and one in mechanical engineering from National Taiwan University of Science and Technology. He was an academic visitor at Bellcore (now Telcordia), USA; University of Oxford, UK; and University of Cambridge, UK. He owns 40 patents and has authored and coauthored more than 250 journal articles and international conference papers in the fields of optics and photonics. His expertise is in fiber-based devices, optical communication, fiber sensing, and reliability testing. Currently, Dr. Liaw is a distinguished professor and vice dean of the National Taiwan University of Science and the secretary-general of the Taiwan Photonic Society. He is a senior member of IPS, OSA, and SPIE.

Contents

Preface   XI

Chapter 1   1
Introductory Chapter: An Overview the Methodologies and Applications
of Fiber Optic Sensing
by Shien-Kuei Liaw

Chapter 2   7
Theoretic Study of Cascaded Fiber Bragg Grating
by Mofreh Toba and Fathy Mohamed Mustafa

Chapter 3   23
Femtosecond Transient Bragg Gratings
by Avishay Shamir, Aviran Halstuch and Amiel A. Ishaaya

Chapter 4   43
Vital Sign Measurement Using FBG Sensor for New Wearable Sensor
Development
by Shouhei Koyama and Hiroaki Ishizawa

Chapter 5   59
The State-of-the-Art of Brillouin Distributed Fiber Sensing
by Cheng Feng, Jaffar Emad Kadum and Thomas Schneider

Preface

In recent years, much effort has been devoted to the study, development, and application of point-to-point fiber sensing for various parameter sensing. Fiber Bragg gratings (FBGs) are key components in the endeavor, usually fabricated using UV laser sources and a phase mask or interferometric techniques. An FBG can be used as a band reject filter; to detect strain, pressure, and temperature; and in telecommunication systems for wavelength selection, among other uses. On the other hand, distributed fiber sensing can monitor the environment along the fiber change based on the Brillouin scattering effect. Distributed Brillouin sensing technique has developed rapidly over the last thirty years. Quite a few investigations on the performance enhancement of Brillouin sensors have been conducted on sensing distance and spatial resolution, paving the way to industrial and commercial applications.

This book presents recent advances in fiber sensing technologies, both in theoretical and real applications, reflecting the cutting-edge technologies and research achievements within these research fields. After a rigorous review process, the editors selected five outstanding chapters from among the submissions for inclusion in this contributor volume. Of these, four are focused on the subject of point-to-point fiber sensing, and the fifth covers distributed fiber sensing. The authors work in academia and industry in Austria, United States, Korea, and Taiwan.

The book consists of the following chapters:

In Chapter 1, "Introductory Chapter: An Overview of the Methodologies and Applications of Fiber Optic Sensing," the editors briefly address the importance of fiber optic sensing, which may be applied in various fields where optical fiber is used either as a transmission medium or as a sensing head. Point-to-point fiber sensing using fiber Bragg gratings (FBGs) and distributed fiber sensing based on Brillouin scattering effect will be introduced. Some prior works based on either of these fiber sensing methodologies are introduced.

In Chapter 2, Fathy Mohamed Mustafa and Mofreh Toba introduce the "Theoretic Study of Cascaded Fiber Bragg Grating." They simulate and analyze the spectral characteristics of the fiber Bragg grating to obtain narrow bandwidth and minimization side lobes in reflectivity. Model equations of cascaded uniform fiber Bragg grating and different cascaded apodization functions are numerically handled and processed via specially cast software to achieve maximum reflectivity, narrow bandwidth without side lobes. For better performance, the proper values for grating length and refractive index modulation must be chosen to achieve maximum reflectivity and narrow bandwidth.

In Chapter 3, "Femtosecond Transient Bragg Gratings" are investigated by Avishay Shamir et al. The authors briefly review the advantages of femtosecond fabrication of fiber Bragg gratings. Then they focus on transient FBGs for optical switching. An experimental result is achieved on generation and characterization of the transient FBGs. A possible mechanism to realize high-power femtosecond laser is introduced. The immunization technique presented here can be used to implement transient thermal gratings in transparent materials and may serve as a diagnostic tool for dielectric materials with different compositions and doping.

In Chapter 4, the "Vital Sign Measurement Using FBG Sensor for New Wearable Sensor," by Shouhei Koyama et al., the authors measured the vital signs from a living body by installing the FBG sensor at a pulsation point. The method of calculating each vital sign from the FBG sensor signal was described. The FBG sensor signal was found to correspond to changes in diameter of the artery caused by the pressure of the blood flow. Vital signs such as pulse rate, respiratory rate, stress load, and blood pressure could be calculated by the FBG sensor head. All the vital signs were calculated with high accuracy. The study helps establish that these vital signs can be calculated continuously and simultaneously.

In Chapter 5, Cheng Feng et al. study "The State-of-the-Art of Brillouin Distributed Fiber Sensing." This chapter provides an overview of different Brillouin sensing techniques and mainly focuses on the most widely used one, the Brillouin optical time domain analysis (BOTDA). The history and development of Brillouin sensing regarding the performance enhancement in various methods and their records will be reviewed, commented on, and compared with one other. In addition, related sensing errors and limitations will be discussed together with corresponding strategies to avoid them.

As editor, I would like to take the opportunity to express my sincere gratitude to all the authors and coauthors who contributed manuscripts to this edited book. We also thank Ms. Jane Liao and Ms. Minglun Tsai for their kind help in typo and format checking.

Dr. Shien-Kuei Liaw
National Taiwan University of Science and Technology,
Taipei City, Taiwan

Introductory Chapter: An Overview the Methodologies and Applications of Fiber Optic Sensing

Shien-Kuei Liaw

1. Introduction

Fiber sensors have several advantages compared to some conventional sensors. They are lightweight and have a small size, high resolution, and good stability; fiber sensors not only are insensitive to electromagnetic interference but also can withstand high temperature and radiation. A variety of linear and nonlinear optical transduction mechanisms have been studied in the last 30–40 years, dealing with the conversion from all kinds of measurands to local measurable optical effects in the fiber. There are previous works, for instance, that designed temperature-compensated fiber Bragg grating (FBG) sensor for monitoring the stress [1], FBG-integrated spherical-shape structure for refractive index sensing [2], and D-shaped fiber combined with a FBG for refractive index and temperature sensing [3]. Fiber sensor can measure and/or monitor many parameters such as strain, weight, temperature, speed, pressure, and so on. Moreover, fiber sensor can also measure the variation of light intensity, wavelength, frequency, phase, and polarization by combining other detectors with optical fiber. Firstly, optical fiber sensors for temperature and pressure have been developed for measurement in oil wells. For example, a precise and real-time ammonia monitoring technique is important especially for gas sensing [3]. Once the gas leakage happens, an immediate alarm is helpful to prevent danger. Secondly, fiber sensing is also used to make a hydrogen sensor. Temperature can be measured by using a fiber that has evanescent loss with various temperature ranges or by analyzing the Brillouin scattering in the optical fiber. Thirdly, angle measurement sensors can be designed based on the Sagnac effect. In recent years, various sensing materials are available for biosensor fabrication, so various fiber-optic biosensors have been proposed and demonstrated. Finally, optical fiber sensors have been developed to simultaneous measurement of temperature and strain with very high accuracy by using fiber Bragg gratings.

2. Types of fiber-optic sensing

2.1 Intrinsic sensing and extrinsic sensing

According to the role optical fiber plays, fiber sensing can be divided into intrinsic sensing and extrinsic sensing. The intrinsic sensing is that the optical fiber itself plays as both the sensing element and the transmission media, as is shown in **Figure 1(a)**;

the extrinsic sensing is that the optical fiber just plays as the transmission media, as is shown in **Figure 1(b)**. Both of them are important and are frequently used in temperature sensing, strain sensing, or pressure sensing, depending on which parameters we want to measure. Specifically, intrinsic fiber-optic sensors provide distributed sensing over a long-distance zone [4], and extrinsic sensors can help us reach inaccessible places, for example, the measurement of temperature inside aircraft jet engines and the measurement of the high temperature inside the electrical transformer. Extrinsic fiber-optic sensors provide excellent protection of measurement signals against noise corruption, and they can be used to measure vibration, rotation, displacement, velocity, acceleration, torque, and temperature [5].

2.2 Types of fiber-optic sensing

Fiber-optic sensors can be split into two big categories: point-by-point sensors and distributed sensors. On the one hand, the point-to-point sensors are usually

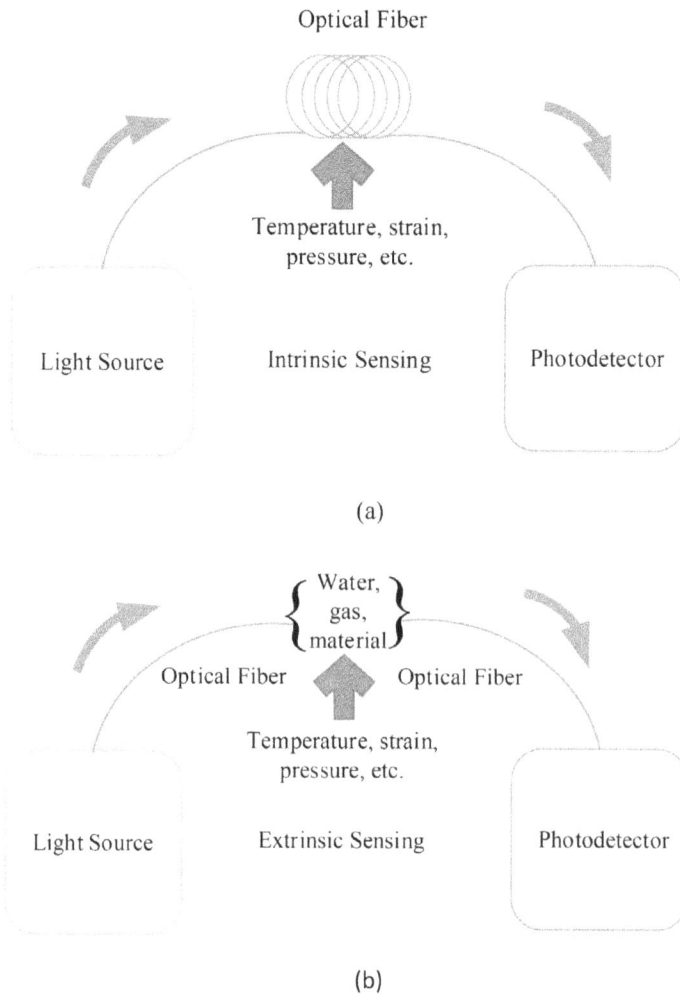

(a)

(b)

Figure 1.
(a) The intrinsic sensing, the optical fiber plays as both the sensing element and the transmission media; and (b) the extrinsic sensing, the optical fiber just plays as the transmission media.

based on FBG. They can measure parameters at a particular location where there is a FBG with high resolution and sensitivity. A standard FBG-based sensing system is shown in **Figure 2** with most required components. Using an optical switch (OSW), a broadband light source may transmit to bridge 1, bridge 2, or bridge 3, respectively, for strain, temperature, and/or stress sensing. Several FBGs are used to monitor multiple parameters/points at the same time. The optical switch is used to share the cost. FBG here is not only the sensing element but also the cavity end of fiber laser. The reflected signals are detected by an optical spectrum analyzer (OSA). Then the data may be in time analyzed by a data logger.

On the other hand, the detectable range of the distributed sensors is based on the Brillouin scattering effect with moderate resolution and limited distances. Nevertheless, a sensor head like a FBG is not required, so distributed sensors are more cost-effective than numerous FBG sensors in long-range sensing distance. A standard Brillouin optical time-domain analysis (BOTDA) sensing system is shown in **Figure 3** with most required components. Firstly, a highly coherent DFB laser source is split into pump source and probe source by using a 50/50 fiber coupler. An erbium-doped fiber amplifier (EDFA1) is used to boost the laser power. A pulse pattern generator is used to drive the electrooptic modulator (EOM1). Then a microwave sine wave around the optical fiber Brillouin frequency of ~11 GHz is fed into the probe source. The signals are scrambled by polarization controllers before they arrive at the Mach-Zehnder modulators (EOMs). The pumped light at the left-hand side is further amplified by an EDFA2 and launched into the fiber under test (FUT) region. The reflected pumped light and probe light comes from other side travel through the optical circulator (OC) and then to EDFA3. A tunable filter or its equivalent is used to filter out pump backscattering and the upper sideband signal. Then the residual signal is monitored and analyzed by a real-time oscilloscope.

In general, point-by-point sensing is practical for short distance and remote monitoring up to 100 km. Distributed sensing based on the Brillouin scattering effect is used to detect strain and temperature for up to 10 km.

2.3 Fiber-optic sensing system

Figure 4 represents a standard fiber-optic sensing system [6]. There is a light source (laser or LED) launching into the optical fiber, and at the right-hand side is

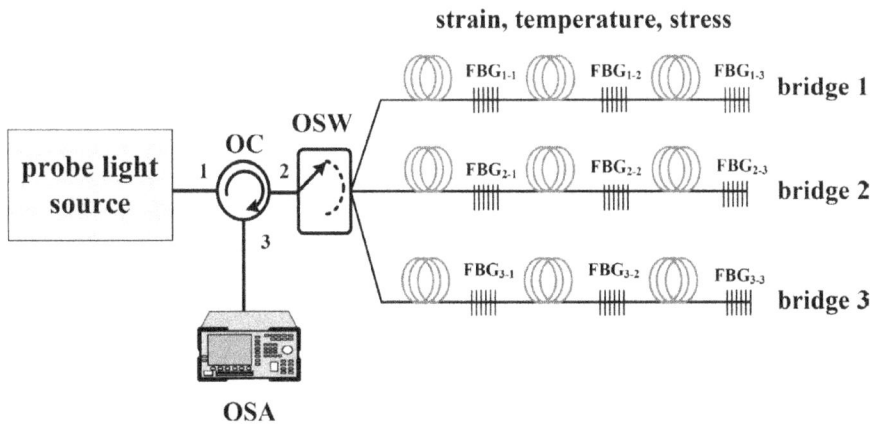

Figure 2.
An example of FBG-based sensing system: OC, optical circulator; OSW, optical switch; OSA, optical spectrum analyzer; FBG, fiber Bragg grating.

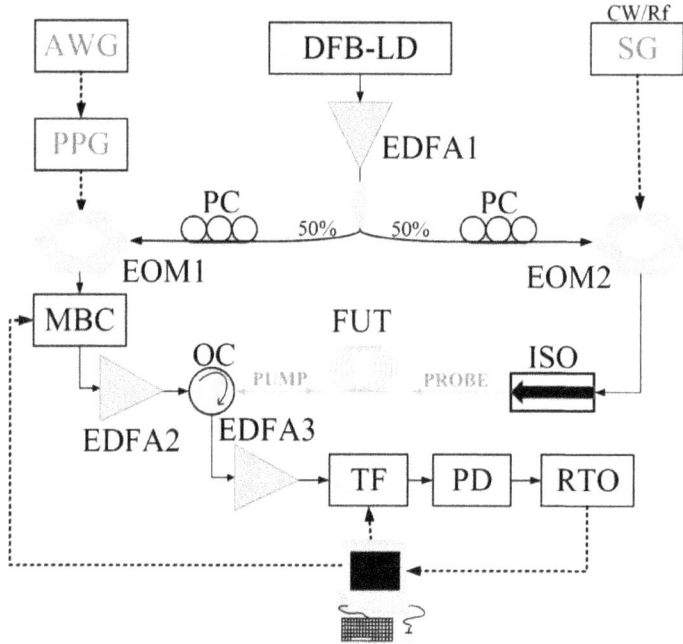

Figure 3.
An example of BOTDA setup: DFB-LD, distributed feedback laser diode; SG, signal generator; EDFA, erbium-doped fiber amplifier; PC, polarization controller; EOM, electrooptical modulator; ISO, optical isolator; AWG, arbitrary waveform generator; PPG, pulse pattern generator; MBC, modulator bias controller; MBC, modulator bias controller; PD, photodetector; RTO, real-time oscilloscopes; OC, optical circulator; FUT, fiber under test.

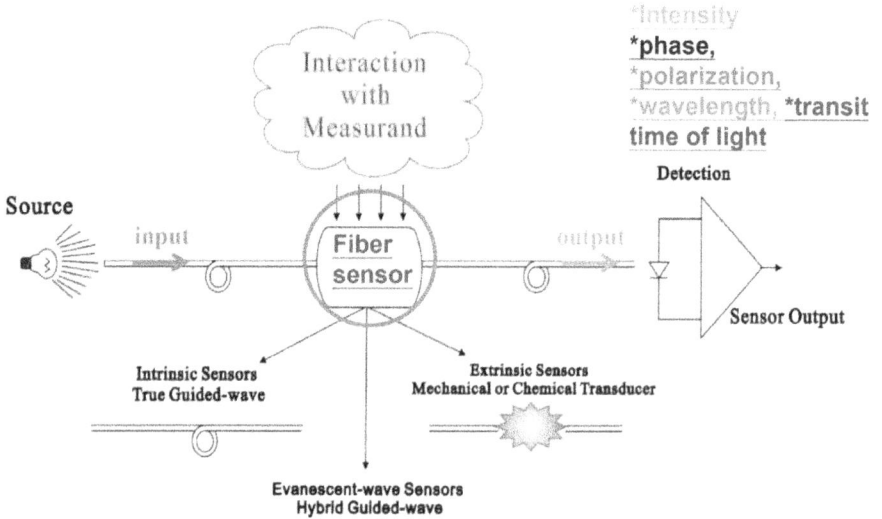

Figure 4.
A standard fiber-optic sensing system [6].

the detector sensing the signal output. The type of fiber sensing may be intrinsic sensing or extrinsic sensing. The parameters such as intensity, phase, polarization, wavelength, and other measurands can be detected and sensed when the light source passes through the monitoring zone where these parameters have direct or indirect effect on the propagating light source.

3. A brief review of previous works

In this section, some previous works of sensing are introduced and addressed. In [7], the authors proposed a FBG liquid level sensor based on the Archimedes' law of buoyancy [8]. They experimentally demonstrated the capability of the proposed device to perform the measurement of water level. It is quite simple to design the device for specific applications without changing the complex cantilever structure. In [9], a D-shaped fiber structure comblined with a FBG for refractive index (RI) and temperature sensing is experimentally investigated. The possibility of simultaneous measurement of the RI and temperature relies on monitoring the resonance dip of the D-shaped fiber modal interferometer and the Bragg wavelength of the FBG. An online monitor of moisture concentration in transformer oil that permits the control of moisture buildup is proposed in [10]. The authors presented a methodology for measurement of moisture concentration in transformer oil using a poly(methyl methacrylate) (PMMA)-based optical FBG. In [11], the authors demonstrated a Brillouin optical correlation domain analysis (BOCDA) system with high-speed random access measurement and temporal gating scheme to extend the range of measurement. Dynamic strain applied at two points was selected arbitrarily along the fiber, and it was measured simultaneously [11]. Other Brillouin scattering effect based fiber sensing has also been addressed in [12]. In summary, fiber sensing is more and more important, and quite a few applications could be found in daily lives.

Author details

Shien-Kuei Liaw
National Taiwan University of Science and Technology, Taiwan (R.O.C.)

*Address all correspondence to: peterskliaw@gmail.com

IntechOpen

© 2019 The Author(s). Licensee IntechOpen. This chapter is distributed under the terms of the Creative Commons Attribution License (http://creativecommons.org/licenses/by/3.0), which permits unrestricted use, distribution, and reproduction in any medium, provided the original work is properly cited.

References

[1] Ren F, Zhang W, Li Y, Lan Y, Xie Y, Dai W. The temperature compensation of FBG sensor for monitoring the stress on hole-edge. IEEE Photonics Journal. 2018;**10**(4). DOI: 10.1109/JPHOT.2018.2858847

[2] Gu M, Yuan S, Yuan Q, Tong Z. Temperature-independent refractive index sensor based on fiber Bragg grating and spherical-shape structure. Optics and Lasers in Engineering. 2019;**115**:86-89

[3] Jian S, Dong Y, Xiao H, Wu B, Xiao S. Refractive index and temperature sensor based on D-shaped Fiber combined with a Fiber Bragg grating. IEEE Sensors Journal. 2018:1-1

[4] Strong AP, Lees G, Hartog AH, Twohig R, Kader K, Hilton G. An integrated system for pipeline condition monitoring. In: International Petroleum Technology Conference; December 2009; Doha, Qatar

[5] Roland U, Renschen CP, Lippik D, Stallmach F. A new fiber optical thermometer and its application for process control in strong electric, magnetic, and electromagnetic fields. Sensor Letters. 2003;**1**(1):93-98

[6] Lee C-L. Optical fiber sensing lecture 2018. Taiwan: National United University

[7] Consales M, Principe S, Iele A, Leone M, Zaraket H, Jomaa I, et al. A fiber Bragg grating liquid level sensor based on the Archimedes' law of buoyancy. Journal of Lightwave Technology. 2018;**36**(20):4936-4941

[8] Keighley HJP. Archimedes' principle and flotation. In: Work Out Physics 'O' Level and GCSE (Macmillan Master Series). London, U.K.: Palgrave. p. 1986

[9] Dong Y, Xiao S, Wu B, Xiao H, Jian S. Refractive index and temperature sensor based on D-shaped Fiber combined with a Fiber Bragg grating. IEEE Sensors Journal. 2019;**19**(4):15

[10] Zhang W, Webb DJ. PMMA based optical fiber Bragg grating for measuring moisture in transformer oil. IEEE Photonics Technology Letters. 2016;**28**(21):2427-2430

[11] Zhang C, Kishi K, Hotate K. Enlargement of measurement range in Brillouin optical correlation domain analysis with high-speed random accessibility using temporal gating scheme for multiple-points dynamic strain measurement. In: Proc. SPIE 9634, 24th International Conference on Optical Fibre Sensors (OFS24); Curitiba, Brazil. 2015

[12] Motil A, Bergman A, Tur M. Invited, State of the art of Brillouin fiber-optic distributed sensing. Optics & Laser Technology. 2016;**78**:81-103

Chapter 2

Theoretic Study of Cascaded Fiber Bragg Grating

Mofreh Toba and Fathy Mohamed Mustafa

Abstract

The purpose of this chapter is to simulate and analyze the spectral characteristics of the fiber Bragg grating (FBG) to obtain narrow bandwidth and minimization side lobes in reflectivity. Fiber Bragg grating has made a big revolution in telecommunication systems. The existence of fiber Bragg grating is needed when an optical fiber amplifier and filter are used. They can be used as band reject filter or band pass filter for optical devices. The model equations of the cascaded uniform fiber Bragg grating and different cascaded apodization functions such as, Hamming apodized fiber Bragg grating, Barthan apodized fiber Bragg grating, Nuttall apodized fiber Bragg grating, Sinc apodized fiber Bragg grating and Proposed apodized fiber Bragg grating are numerically handled and processed via specially cast software to achieve maximum reflectivity, narrow bandwidth without side lobes.

Keywords: fiber Bragg grating (FBG), reflectivity, grating length and narrow bandwidth

1. Introduction

FBGs are typically used as a selective wave-length reflector. Fiber Bragg grating nuts are spectral filters based on the Bragg reflection principle. The light usually reflects the narrow wavelength and sends all other wavelengths. When light is spread by periodic rotation of regions of the upper and lower refractive index, it is partially reflected in each interface between those regions [1]. The power of coupling, and hence the reflection and transmission spectra at an angle of inclination, fiber geometry, and the refractive index (RI) of the surrounding medium are affected [2]. There are a number of parameters in which FBG spectra have been shown, such as change in refractive index, bending of fibers, period of grating, excitation conditions, temperature, and length of tree fibers [3]. The fiber Bragg grating separator (FBG) is an optical device that periodically changes the refractive index along the direction of propagation at the heart of the fiber. The basic property of FBGs is that they reflect the light in a narrow band centered around Bragg wave length. There is a different structure of FBG such as uniform, wet, peep, slanted and long period. When light diffuses through FBG in a narrow band of wavelength, the total reflection occurs at the Bragg wavelength and the other wavelength is not affected by the Bragg derivation except for some side lobes present in the reflection spectrum. These side lobes can be suppressed using the coding technique. The reflection range depends on the length and force of the refractive index formation. Wave reflection also depends on temperature and voltage [4]. In order to achieve

IntechOpen

high-efficiency long-range fiber connections, WDM is introduced. Scattering is a key factor limiting the design of long-distance optical links. Several techniques have achieved effective dispersion compensation (DC). The widely used technologies are DCF and broken glass panels (CFBG). Although DCF is a large-scale unit, CFBG is superior to many faces [5]. Due to its excellent multicast capabilities, the fiber Bragg grating (FBG) sensors are particularly attractive for applications where a large number of sensors are desirable such as industrial process control, fire detection systems, and temperature conversion of power, since sensor FBG occupies a narrowband bandwidth that is very narrow and can easily create a distributed sensor matrix by writing several FBG sensors on a single fiber at different locations [6]. The refractive index of the nucleus is permanently changed. Germanium doped silica fiber is used in the manufacture of FBG because it is sensitive which means that the refractive index of the nucleus changes through exposure to light. The amount of change depends on the intensity and duration of exposure. It also depends on optical fiber sensitivity, so the level of fission with germanium should be high for high reflectivity [7]. Fiber Bragg grating nuts are spectral filters based on the Bragg reflection principle. The light usually reflects the narrow wavelength and sends all other wavelengths. When light is spread by periodic rotation of regions of the upper and lower refractive index, it is partially reflected in each interface between those regions [8]. This chapter is organized as follows. After the introduction in Section 1, Section 2 presents a basic model and analysis. In Section 3, we present the proposed system. The simulation results are displayed and discussed in Section 4. Finally we devoted to the main conclusions.

2. Basic model and analysis

In the present section, the basic model, governing equations and the analysis of the fiber Bragg grating are investigated to obtain a maximum reflectivity and minimum bandwidth, we discuss more than one case of different models of fiber Bragg grating; we will discuss in this section two models of fiber Bragg grating:

1. Uniform fiber Bragg grating

2. Apodized fiber Bragg grating

2.1 Uniform fiber Bragg grating

2.1.1 Fiber Bragg grating structure

The basic structure of the uniform fiber Bragg grating is illustrated in **Figure 1** [7, 9]. As shown in **Figure 1**, the refractive index of the core is modulated by a period of Λ. When light is transmitted through the fiber which contains a segment of FBG, part of the light will be reflected. The reflected light has a wavelength equals to the Bragg wavelength so that it is reflected back to the input while others are transmitted. The term uniform means that the grating period, Λ, and the refractive index modulation, δn, are constant over the length of the grating. A grating is a device that periodically modifies the phase or the intensity of a wave reflected on, or transmitted through it [10]. The equation relating the grating spatial periodicity and the Bragg resonance wavelength is given by $\lambda_B = 2n_{eff}\Lambda$. Where n_{eff} the effective mode is index and Λ is the grating period [11].

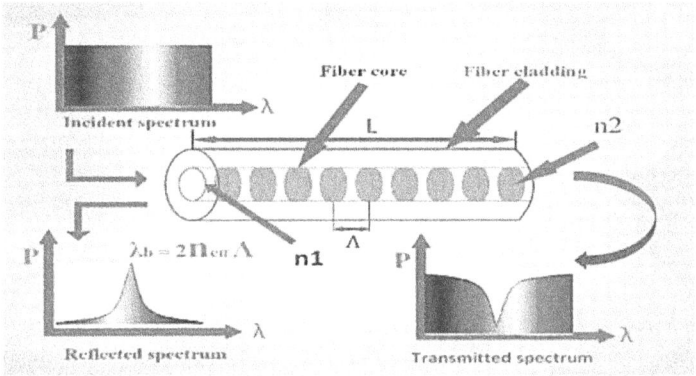

Figure 1.
Basic structure of fiber Bragg grating.

2.1.2 Theory and principle of operation

The study of the spectral characteristics of the uniform fiber Bragg grating is carried out by solving the dual mode equations. Dual mode theory is an important tool for understanding the design of fiber dividers from fiber Bragg grating [7]. FBG can be considered as a weak wave structure so that the pair mode theory can be used to analyze light propagation in a weak waveguide structure such as FBG. Dual-mode equations that describe the propagation of light can be obtained in FBG using the couple mode theory. The theory of marital status was first introduced in the early 1950s to microwave devices and later applied to optical devices in early 1970 [11].

For maximum reflectivity [12]:

$$R_{peak} = \tanh^2(kL) \qquad (1)$$

Reflective bandwidth, $\Delta\lambda$ of uniform FBG is defined as wavelength bandwidth between the first zero reflective wavelength of both sides of peak reflection wavelength. It can be calculated by a general expression of the approximate bandwidth of the grating is:

$$\Delta\lambda = \lambda_{B\,s}\sqrt{\left(\frac{\Delta n_{ac}}{2n_{eff}}\right)^2 + \left(\frac{1}{N}\right)^2} \qquad (2)$$

Where λ_B is the Bragg (center) wavelength, s is a parameter indicting the strength of the gratings (~1 for strong gratings and ~0.5 for weak gratings), N is the number of grating planes, Δn_{ac} is the change in the refractive index and n_{eff} is the effective refractive index.

The forward propagated light is reflected at Bragg wavelength [13]:

$$\lambda_B = 2n\Lambda \qquad (3)$$

Where λ_B is the Bragg wavelength (wavelength of the reflection peak amplitude), n is the effective refractive index of optical mode propagating along the fiber and Λ is the period of FBG structure. For a uniform Bragg grating formed within the core of an optical fiber with an average refractive index n_o. The index of the refractive profile can be expressed as [6, 14]:

$$n(z) = n_0 + \Delta n \cos\left(\frac{2\pi z}{\Lambda}\right) \tag{4}$$

Where Δn is the amplitude of the induced refractive index perturbation, Λ is the nominal grating period and z is the distance along the fiber longitudinal axis. Using coupled-mode theory the reflectivity of a grating with constant modulation amplitude and period is given by the following expression [6, 8, 12]:

$$R(l,\lambda) = \frac{k^2 \sinh^2(sl)}{\Delta\beta^2 \sinh^2(sl) + s^2 \cosh^2(sl)} \tag{5}$$

Where $R(l,\lambda)$ is the reflectivity, which is a function of the grating length L and wavelength λ, $\Delta\beta = \beta - \pi/\Lambda$ is the detuning wave vector, $\beta = 2\pi n_0/\lambda$ is the propagation constant and $s^2 = k^2 - \Delta\beta^2$ and $\kappa = \frac{\pi\Delta n}{\lambda} Mpower$.

$Mpower$ is ac coupling coefficient, $Mpower$ is the fraction of the fiber mode power contained by the fiber core.

In the case where the grating is uniformly written through the core, $Mpower$ can be approximated by, $1 - V^2$, where $V = \frac{2\pi}{\lambda} a \sqrt{n^2_{c0} - n^2_{cl}}$ is the normalized frequency of the fiber, a is the core radius, n_{CO} and n_{CL} are the core and cladding indices, respectively. At the center wavelength of the Bragg grating the wave vector detuning is $\Delta\beta = 0$, therefore the expression for the reflectivity becomes:

$$R(l,\lambda) = \tanh^2(kl) \tag{6}$$

The reflectivity increases as the induced index of refraction change gets larger. Similarly, as the length of the grating increases, so does the resultant reflectivity.

2.2 Apodized fiber Bragg grating

In the present section, we cast the basic model and the governing equation to apodized fiber Bragg grating. Apodized FBG offer significant improvement in side lobe suppression but on the expense of reducing the peak reflectivity. Apodized gratings have variations along the fiber in the refractive index modulation envelope ($\Delta n\alpha c$) with constant grating period and constant DC refractive index function. The index of the refractive profile of Apodized can be expressed as [6]:

$$n(z) = n_{c0} + \Delta n_0 A(z)n_d(z) \tag{7}$$

Where n_{c0} is the core refractive index, Δn_0 is the maximum index variation, $n(z)$ is the index variation function and (z) is the Apodization function. Apodization profiles are [6, 13, 15]:

1. Uniform:

$$A(z) = 1, \, 0 \leq z \leq L \tag{8}$$

2. Hamming Function

$$A(z) = 0.54 - 0.46 \cos\left(\frac{2\pi z}{L}\right), \text{ where } 0 \leq z \leq L \tag{9}$$

3. Barthan Function:

$$A(z) = 0.62 - 0.48 \left|\frac{z}{L} - 0.5\right| + 0.38 \cos\left(\frac{z}{L} - 0.5\right), \text{ where, } 0 \leq z \leq L \tag{10}$$

4. Nuttall Function:

$$A(z) = 0.3635819 - 0.48917755 \left(2\pi\frac{z}{L}\right) + 0.1365996 \left(4\pi\frac{z}{L}\right) - 0.0106411\left(6\pi\frac{z}{L}\right),$$

(11)

where, $0 \leq z \leq L$

5. Sinc Function:

$$A(z) = \text{sinc}\left(2\pi\frac{z - \frac{L}{2}}{L}\right), 0 \leq z \leq L$$

(12)

6. Proposed ($\cos^8$) Function [8]:

$$A(z) = \left(\cos\left(\frac{2z}{L} - 1\right)\right)^8, 0 \leq z \leq L$$

(13)

3. The proposed system

This section shows a proposed model for cascaded n stages of FBGs. Analysis of this model is done by coupling theory [16]. T matrix 2×2 where FBG is divided into sections.

Each section is shown in **Figure 2** where T is the length of each section and Λ is the space between reflected planes of each section where:

a_0: is the incident optical signal.
b_0: is the reflected optical signal.
a_m: is the output optical signal.
b_m: is the reflected optical signal at output of grating.
m: is number of sections.

The transfer matrix can be expressed by small multiplied matrixes as in:

$$\begin{bmatrix} a_0 \\ b_0 \end{bmatrix} = [T^1][T^2][T^3][T^4]...[T^m]\begin{bmatrix} a_m \\ b_m \end{bmatrix}$$

(14)

Replacing m matrix by whole matrix [**T**] where:

$$[\mathbf{T}] = \prod_{j=1}^{m}[T^m]$$

(15)

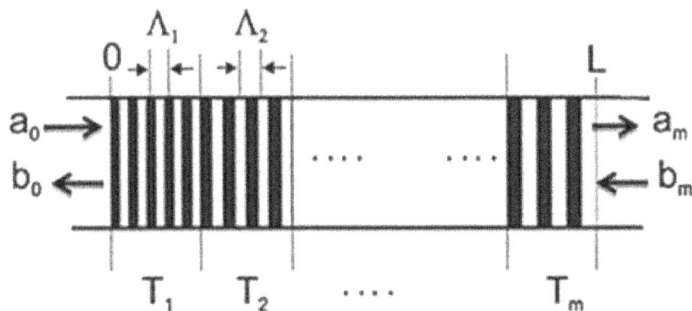

Figure 2.
Fiber Bragg grating sections.

The transfer matrix can be written as:

$$\begin{bmatrix} a_0 \\ b_0 \end{bmatrix} = [\mathbf{T}] \begin{bmatrix} a_m \\ b_m \end{bmatrix} \tag{16}$$

Where $[\mathbf{T}]$ can be written as:

$$\begin{bmatrix} a_0 \\ b_0 \end{bmatrix} = \begin{bmatrix} T_{11} & T_{12} \\ T_{21} & T_{22} \end{bmatrix} \begin{bmatrix} a_m \\ b_m \end{bmatrix} \tag{17}$$

In case of FBG reflection, There is no reflection at o/p at distance L so $b_m = 0$

3.1 In case of one grating

Transfer matrix can be written as:

$$\begin{bmatrix} a_0 \\ b_0 \end{bmatrix} = \begin{bmatrix} T_{11} & T_{12} \\ T_{21} & T_{22} \end{bmatrix} \begin{bmatrix} a_m \\ 0 \end{bmatrix} \tag{18}$$

Where Transfer matrix parameters T_{11} and T_{21} are:

$$a_0 = T_{11} a_m \tag{19}$$

$$b_0 = T_{21} a_m = \frac{a_0 T_{21}}{T_{11}} \tag{20}$$

Then reflectivity R can be calculated by:

$$R = |\rho_1|^2 \tag{21}$$

Where,

$$\rho_1 = \frac{b_0}{a_0} = \frac{T_{21}}{T_{11}} \tag{22}$$

3.2 In case of two cascaded gratings

Figure 3 shows the connection between two FBGs where the output of the first on is connected to input of the second. In this case, the input optical signal for the second stage of the grating from (7) is:

$$b_0 = a_{02} = a_{01} \frac{T_{21}}{T_{11}} \tag{23}$$

Transfer matrix can be written as:

$$\begin{bmatrix} a_{01} \dfrac{T_{21}}{T_{11}} \\ b_{02} \end{bmatrix} = \begin{bmatrix} T_{11} & T_{12} \\ T_{21} & T_{22} \end{bmatrix} \begin{bmatrix} a_m \\ 0 \end{bmatrix} \tag{24}$$

Transfer matrix parameters T_{11} and T_{21} in two cascaded are:

$$a_{01} \frac{T_{21}}{T_{11}} = T_{11} a_m \tag{25}$$

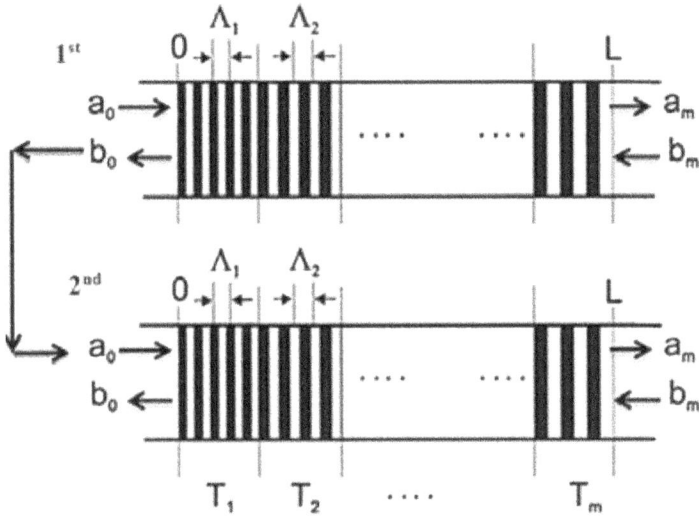

Figure 3.
Connection between two cascaded FBG.

Then from Eq. (12) we get:

$$a_{01} = \frac{T_{11}^2}{T_{21}} a_m \tag{26}$$

$$b_{02} = T_{21} a_m \tag{27}$$

We can calculate Reflectivity for the second grating by:

$$R = |\rho_2|^2 \tag{28}$$

Where,

$$\rho_2 = \frac{b_{02}}{a_{01}} = \frac{T_{21} * a_m * T_{21}}{T_{11}^2 * a_m} \tag{29}$$

$$\rho_2 = \frac{T_{21}^2}{T_{11}^2} \tag{30}$$

$$\therefore \rho_2 = (\rho_1)^2 \tag{31}$$

3.3 In case of third cascaded gratings

The output of second grating b_{02} is equal to input of third grating:

$$a_{03} = b_{02} = a_{01} \frac{T_{21}^2}{T_{11}^2} \tag{32}$$

Transfer matrix can be written as:

$$\begin{bmatrix} a_{01}\dfrac{T_{21}^2}{T_{11}^2} \\ b_{03} \end{bmatrix} = \begin{bmatrix} T_{11} & T_{12} \\ T_{21} & T_{22} \end{bmatrix} \begin{bmatrix} a_m \\ 0 \end{bmatrix} \tag{33}$$

Transfer matrix parameters T_{11} and T_{21} in two cascaded are:

$$a_{01} \frac{T_{21}^2}{T_{11}^2} = T_{11} a_m \tag{34}$$

$$a_{01} = \frac{T_{11}^3}{T_{21}^2} a_m \tag{35}$$

$$b_{03} = T_{21} a_m \tag{36}$$

We can calculate reflectivity for the third grating by:

$$R = |\rho_3|^2 \tag{37}$$

Where,

$$\rho_3 = \frac{b_{03}}{a_{01}} = \frac{T_{21} * a_m * T_{21}^2}{T_{11}^3 * a_m} \tag{38}$$

$$\rho_3 = \frac{T_{21}^3}{T_{11}^3} \tag{39}$$

$$\therefore \rho_3 = (\rho_1)^3 \tag{40}$$

From (23) we can prove that the reflectivity of three cascaded FBG at the same parameters is equal to cubic reflectivity of the first one. At n stages of cascaded FBGs have the same parameters and each of them have a reflectivity R; the reflectivity of all n groups is equal to R^n.

4. Simulation results and discussion

In this section we will display the simulation results of the cascaded uniform and cascaded different apodized fiber Bragg grating to obtain narrow bandwidth without side lobes and maximum reflectivity.

4.1 Cascaded uniform fiber Bragg grating

We will simulate the spectral characteristics of the cascaded uniform fiber Bragg grating as in **Figure 4**. In this simulation the modulation index, $d_n = 0.0003$ and grating length L = 5 mm. From **Figure 4** we noted that as the number of cascade of fiber Bragg grating is increased the bandwidth is decreased and the side lobes are also decreased but reflectivity is decreased.

We have obtained from simulation result for one unit of fiber Bragg Grating the reflectivity, R = 99% and bandwidth =0.23 nm but the side lobes is high, for two cascaded units from fiber Bragg grating the reflectivity, R = 98% and bandwidth =0.18 nm but side lobes in this case is decreased on one unit of fiber Bragg grating, for three cascaded units from fiber Bragg grating the reflectivity, R = 97% and bandwidth =0.17 nm but side lobes is decreased and for four cascaded units from fiber Bragg grating the reflectivity, R = 96%, bandwidth =0.16 nm and approximately no side lobes. Then we concluded that Reflectivity, R = 96%, bandwidth = 0.16 nm and the minimum side lobes is achieved at the fourth unit of cascaded fiber Bragg grating.

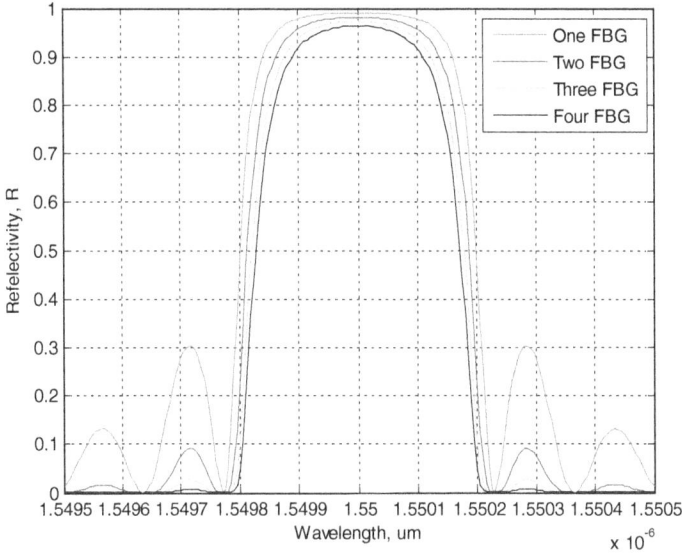

Figure 4.
Reflectivity spectrum for four stage uniform fiber Bragg grating.

4.2 Hamming apodized cascaded fiber Bragg grating

We will simulate the spectral characteristics of the cascaded Hamming apodized fiber Bragg grating as in **Figure 5**.

In this simulation the modulation index, $d_n = 0.0003$ and grating length L = 5 mm. From **Figure 5** we noted that as the number of cascade of fiber Bragg grating is increased the bandwidth is decreased and the side lobes are also decreased but reflectivity is decreased.

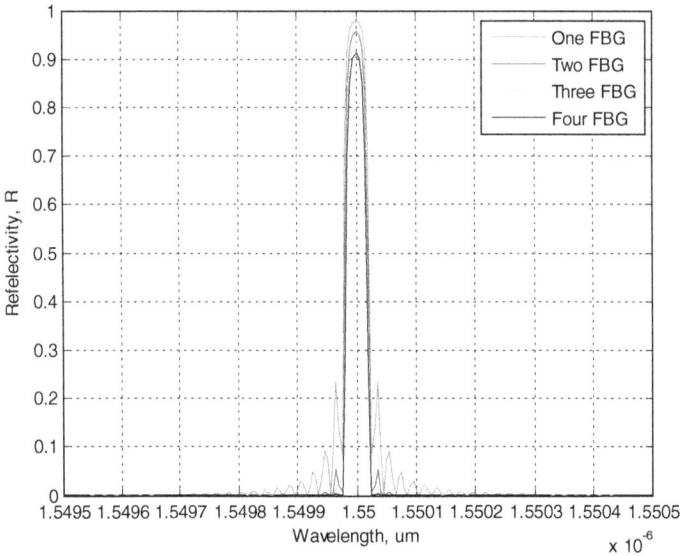

Figure 5.
Reflectivity spectrum for four stage hamming apodized fiber Bragg grating.

We have obtained from simulation result for one unit of fiber Bragg grating the reflectivity, R = 98% and bandwidth =0.026 nm but the side lobes is high, for two cascaded units from fiber Bragg grating the reflectivity, R = 96% and bandwidth =0.020 nm but side lobes in this case is decreased on one unit of fiber Bragg grating, for three cascaded units from fiber Bragg grating the reflectivity, R = 93% and bandwidth =0.018 nm but side lobes is decreased and for four cascaded units from fiber Bragg grating the reflectivity, R = 92%, bandwidth =0.017 nm and approximately no side lobes. Then we concluded that Reflectivity, R = 92%, bandwidth =0.017 nm and the minimum side lobes is achieved at the fourth unit of cascaded fiber Bragg grating.

4.3 Barthan apodized cascaded fiber Bragg grating

We will simulate the spectral characteristics of the cascaded Barthan apodized fiber Bragg grating as in **Figure 6**.

In this simulation the modulation index, $d_n = 0.0003$ and grating length L = 5 mm. From **Figure 6** we noted that as the number of cascade of fiber Bragg grating is increased the bandwidth is decreased and the side lobes are also decreased but reflectivity is decreased. We have obtained from simulation result for one unit of fiber Bragg Grating the reflectivity, R = 100% and bandwidth = 0.083 nm but the side lobes is high, for two cascaded units from fiber Bragg grating the reflectivity, R = 99% and bandwidth =0.069 nm but side lobes in this case is decreased on one unit of fiber Bragg grating, for three cascaded units from fiber Bragg grating the reflectivity, R = 99% and bandwidth =0.066 nm but side lobes is decreased and for four cascaded units from fiber Bragg grating the reflectivity, R = 99%, bandwidth = 0.064 nm and approximately no side lobes. Then we concluded that Reflectivity, R = 99%, bandwidth =0.064 nm and the minimum side lobes is achieved at the fourth unit of cascaded fiber Bragg grating.

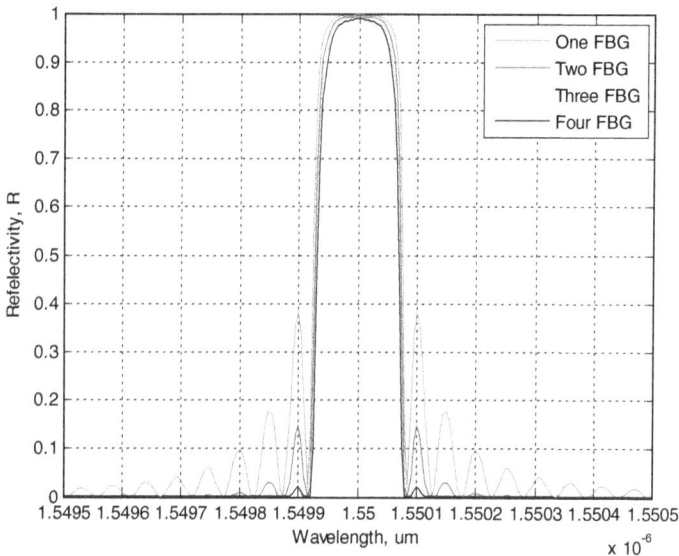

Figure 6.
Reflectivity spectrum for four stage Barthan apodized fiber Bragg grating.

4.4 Nuttall apodized cascaded fiber Bragg grating

We will simulate the spectral characteristics of the cascaded Nuttall apodized fiber Bragg grating as in **Figure 7**.

In this simulation the modulation index, $d_n = 0.0003$ and grating length L = 5 mm. from **Figure 7** we noted that as the number of cascade of fiber Bragg grating is increased the bandwidth is decreased and the side lobes are also decreased but reflectivity is decreased. We have obtained from simulation result for one unit of fiber Bragg grating the reflectivity, R = 99% and bandwidth = 0.08 nm but the side lobes is high, for two cascaded units from fiber Bragg grating the reflectivity, R = 99% and bandwidth =0.063 nm but side lobes in this case is decreased on one unit of fiber Bragg grating, for three cascaded units from fiber Bragg grating the reflectivity, R = 98% and bandwidth = 0.061 nm but side lobes is decreased and for four cascaded units from fiber Bragg grating the reflectivity, R = 98%, bandwidth = 0.059 nm and approximately no side lobes. Then we concluded that Reflectivity, R = 98%, bandwidth = 0.059 nm and the minimum side lobes is achieved at the fourth unit of cascaded fiber Bragg grating.

4.5 Sinc apodized cascaded fiber Bragg grating

We will simulate the spectral characteristics of the cascaded Sinc apodized fiber Bragg grating as in **Figure 8**. In this simulation we choose modulation index, $d_n = 0.0003$ and grating length L = 5 mm. From **Figure 8** we noted that as the number of cascade of fiber Bragg grating is increased the bandwidth is decreased and the side lobes are also decreased but reflectivity is decreased.

we have obtained from simulation result for one unit of fiber Bragg grating the reflectivity, R = 99% and bandwidth =0.013 nm but the side lobes is high, for two cascaded units from fiber Bragg grating the reflectivity, R = 97% and bandwidth =0.01 nm but side lobes in this case is decreased on one unit of fiber Bragg grating, for three cascaded units from fiber Bragg grating the reflectivity, R = 96% and bandwidth =0.0098 nm but side lobes is decreased and for four cascaded units from

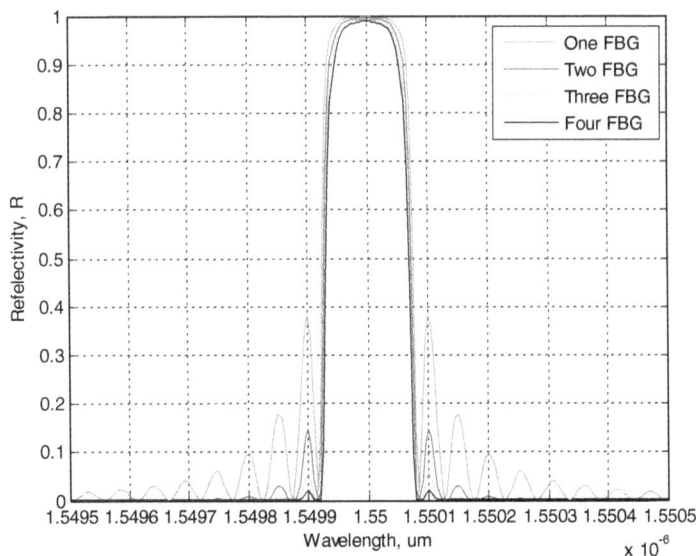

Figure 7.
Reflectivity spectrum for four stage Nuttall apodized fiber Bragg grating.

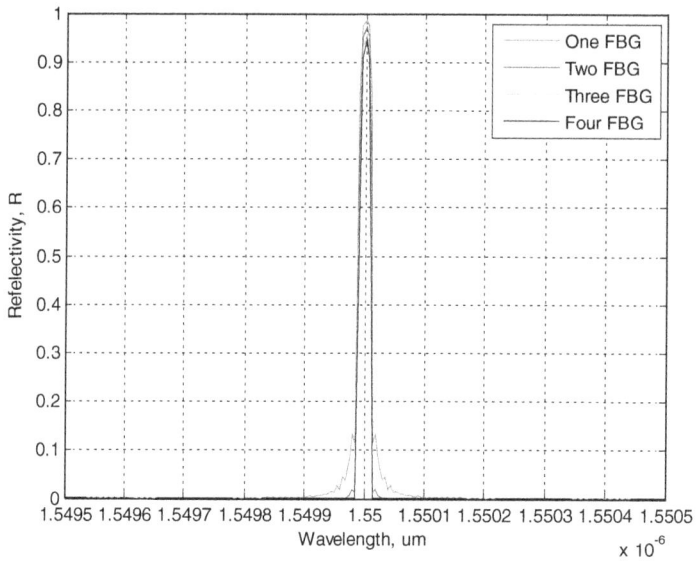

Figure 8.
Reflectivity spectrum for four stage sinc apodized fiber Bragg grating.

fiber Bragg grating the reflectivity, R = 95%, bandwidth =0.0096 nm and approximately no side lobes. Then we concluded that Reflectivity, R = 95%, bandwidth =0.0096 nm and the minimum side lobes is achieved at the fourth unit of cascaded fiber Bragg grating.

4.6 Proposed apodized cascaded fiber Bragg grating

We will simulate the spectral characteristics of the cascaded proposed apodized fiber Bragg grating as in **Figure 9**.

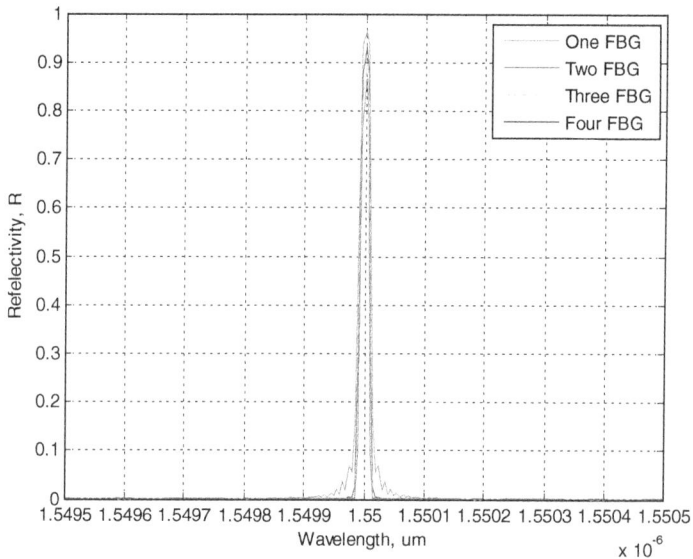

Figure 9.
Reflectivity spectrum for four stage proposed apodized fiber Bragg grating.

In this simulation the modulation index, $d_n = 0.0003$ and grating length L = 5 mm. From **Figure 9** we noted that as the number of cascade of fiber Bragg grating is increased the bandwidth is decreased and the side lobes are also decreased but reflectivity is decreased. We have obtained from simulation result for one unit of fiber Bragg Grating the reflectivity, R = 96% and bandwidth =0.013 nm but the side lobes is high, for two cascaded units from fiber Bragg grating the reflectivity, R = 93% and bandwidth =0.0092 nm but side lobes in this case is decreased on one unit of fiber Bragg grating, for three cascaded units from fiber Bragg grating the reflectivity, R = 90% and bandwidth =0.0084 nm but side lobes is decreased and for four cascaded units from fiber Bragg grating the reflectivity, R = 87%, bandwidth =0.0081 nm and approximately no side lobes. Then we concluded that Reflectivity, R = 87%, bandwidth =0.0081 nm and the minimum side lobes is achieved at the fourth unit of cascaded fiber Bragg grating.

Apodization profile	Grating length and refractive index modulation	Stage number	Reflectivity, R (%)	Bandwidth (nm)
Uniform	L = 5 mm, dn = 0.0003	1st stage	99	0.22
		2nd stage	98	0.17
		3rd stage	97	0.168
		4th stage	96	0.16
Hamming	L = 40 mm, dn = 0.0004	1st stage	98	0.026
		2nd stage	96	0.02
		3rd stage	93	0.018
		4th stage	92	0.017
Barthan	L = 15 mm, dn = 0.0001	1st stage	100	0.083
		2nd stage	99	0.069
		3rd stage	99	0.066
		4th stage	99	0.064
Nuttall	L = 15 mm, dn = 0.0003	1st stage	99	0.08
		2nd stage	99	0.063
		3rd stage	98	0.061
		4th stage	98	0.059
Sinc	L = 80 mm, dn = 0.0004	1st stage	99	0.013
		2nd stage	97	0.010
		3rd stage	96	0.0098
		4th stage	95	0.0096
Proposed	L = 80 mm, dn = 0.0002	1st stage	96	0.013
		2nd stage	93	0.0092
		3rd stage	90	0.0084
		4th stage	87	0.0081

Table 1.
Comparison between reflectivity, R and bandwidth for different cascaded units of apodized fiber Bragg grating.

4.7 Comparisons between reflectivity and bandwidth for different cascaded units of apodized fiber Bragg grating

Table 1 show different cascaded apodized fiber Bragg grating variations of reflectivity, R and bandwidth with the increase of number of stages.

From **Table 1** we noted that as the number of cascaded units of fiber Bragg grating is increased the reflectivity is slightly decreased and the bandwidth decreased with minimum side lobes.

5. Conclusions

In this work the model equations of the cascaded uniform fiber Bragg grating and different cascaded apodization functions are numerically handled and processed via specially cast software to achieve maximum reflectivity, narrow bandwidth and minimum side lobes. For better performance the proper values for grating length and refractive index modulation must be chosen to achieve maximum reflectivity and narrow bandwidth. The minimization in side lobes achieved by using cascaded units from FBG. From this study we concluded that:

1. Uniform FBG in case one unit R = 99%, BW = 0.22 nm and exist side lobes but for fourth unit R = 96%, BW = 0.16 nm with minimum side lobes.

2. For Hamming, Barthan and Nuttall FGB achieved high reflectivity and narrow bandwidth with minimum side lobes in fourth unit FBG.

3. For sinc and proposed FBG achieved narrow bandwidth without side lobes.

4. High reflectivity achieved in case of Barthan apodization function where, R = 99% in fourth unit.

5. Narrow bandwidth achieved in case of proposed apodization function where, BW = 0.0081 nm in fourth unit without side lobes.

Acknowledgements

I always feel indebted to ALLAH whose blessings on me cannot be counted. I would like to thank the anonymous reviewers of *IntechOpen* and *Academic Editor: Dr. Shien-Kuei Liaw* for their valuable comments. Finally, but most importantly, I am really indebted to my parents, sisters, brother and my wife for their continuous support throughout all my studies. Hope they find in my current achievement a reward for their care and love.

Acronyms and abbreviations

FBG   fiber Bragg grating
RI   refractive index
DC   dispersion compensation

Author details

Mofreh Toba[1] and Fathy Mohamed Mustafa[2*]

1 Electrical Engineering Department, Faculty of Engineering, Fayoum University, Fayoum, Egypt

2 Electrical Engineering Department, Faculty of Engineering, Beni-Suef University, Beni-Suef, Egypt

*Address all correspondence to: fmmg80@eng.bsu.edu.eg

IntechOpen

© 2019 The Author(s). Licensee IntechOpen. This chapter is distributed under the terms of the Creative Commons Attribution License (http://creativecommons.org/licenses/by/3.0), which permits unrestricted use, distribution, and reproduction in any medium, provided the original work is properly cited. (cc) BY

References

[1] Gumasta RK, Khare A. Effect of length and apodization on fiber Bragg grating characteristics. International Journal of Scientific & Engineering Research. 2014;5:893-895

[2] Elzahaby EA, Kandas I, Aly MH, Mahmoud K. Sensitivity improvement of reflective tilted FBGs. Applied Optics. 2016;55(12):3306-3312

[3] Kaur R, Bhamrah MS. Effect of grating length on reflection spectra of uniform fiber Bragg gratings. International Journal of Information and Telecommunication Technology. 2011;3(2)

[4] Ghosh C, Alfred QM, Ghosh B. Spectral characteristics of uniform fiber Bragg grating with different grating length and refractive index variation. International Journal of Innovative Research in Computer and Communication Engineering. 2015;3: 456-462

[5] Mohammed N, Okasha MN, Aly HM. A wideband apodized FBG dispersion compensator in long haul WDM systems. Journal of Optoelectronics and Advanced Materials. 2016;18:475-479

[6] El-gammal HM, Fayed HA, Abd El-aziz A, Aly MH. Performance analysis & comparative study of uniform, apodized and pi-phase shifted FBGs for array of high performance temperature sensors. Optoelectronics and Advanced Materials-Rapid Communications. 2015; 9(9):1251-1259

[7] Nagwan IT, Eldeeb WS, El_Mashade MB, Abdelnaiem AE. Optimization of uniform Fiber Bragg grating reflection spectra for maximum reflectivity and narrow bandwidth. International Journal of Computational Engineering Research (IJCER). 2015;5:53-61

[8] Ugale S, Mishra V. Fiber Bragg grating modeling, characterization and optimization with different index profiles. International Journal of Engineering Science and Technology. 2010;2(9):4463-4468

[9] Mahapatra JR, Chattopadhyay M. Spectral characteristics of uniform fiber Bragg grating using couple mode theory. International Journal of Electrical, Electronics and Data Communication. 2013;1:40-44

[10] Arora D, Prakash J, Singh H, Wason A. Reflectivity and Braggs wavelength in FBG. International Journal of Engineering (IJE). 2011;5:341-349

[11] Mahanta DK. Design of uniform fiber Bragg grating using transfer matrix method. International Journal of Computational Engineering Research (IJCER). 2013;03:8-13

[12] Abdallah I, Rachida H, Mohamed C-B. Uniform fiber Bragg grating modeling and simulation used matrix transfer method. IJCSI International Journal of Computer Science Issues. 2012;9(1, 2):368-374

[13] Nazmi AM, Ali TA, Aly MH. Performance optimization of apodized FBG-based temperature sensors in single and quasi-distributed DWDM systems with new and different apodization profiles. AIP Advances. 2013;3:1-21

[14] Turan Erdogan T. Fiber grating spectra. Journal of Lightwave Technology. 1997;15:1277-1294

[15] Mohammed NA, Elashmawy AW, Aly MH. Distributed feedback fiber filter based on apodized fiber Bragg grating. Optoelectronics and Advanced Materials—Rapid Communications. 2015;9(9–10):1093-1099

[16] Kashyap R. Fiber Bragg Grating. San Diego, USA: Academic Press; 2009. p. 316

Chapter 3

Femtosecond Transient Bragg Gratings

Avishay Shamir, Aviran Halstuch and Amiel A. Ishaaya

Abstract

Fiber Bragg gratings (FBGs) have found numerous applications in fiber lasers, sensors, telecommunication, and many other fields. Traditionally, they are fabricated using UV laser sources and a phase mask or other interferometric techniques. In the past two decades, FBGs have been fabricated with femtosecond lasers in either the point-by-point method or by using a phase mask, in a similar configuration as with UV laser sources. In the following, we briefly review the advantages of femtosecond fabrication of fiber Bragg gratings. We then focus on transient FBGs; these are FBGs that exist for a short duration only, for the purpose of all-optical, in-fiber switching and modulation and the possible mechanism to implement them with a high-power femtosecond laser. The theory behind transient grating switching is outlined, and we discuss related experimental results achieved by our group on both permanent grating inscription and the generation of transient (dynamic) fiber Braggs gratings.

Keywords: femtosecond fiber Bragg gratings, transient fiber Bragg gratings, dynamic fiber Bragg grating, all-optical switching and modulation

1. Introduction

Femtosecond laser micromachining and inscription have attracted significant attention in the past decade, not only for material processing applications, such as cutting or drilling [1, 2], but also for the fabrication of 3D photonic devices in transparent materials. When focusing a high-power femtosecond pulse inside a transparent dielectric material, the intensity at the focal region is high enough to initiate multiphoton ionization, which eventually leads to structural changes and permanent refractive index changes [3–7]. This technique has some advantages over current photonic device fabrication methods: (i) the nonlinear nature of the laser-matter interaction confines any induced index change to the focal volume, enabling 3D fabrication of photonic devices in a relatively short time compared to planar semiconductor-based fabrication methods; (ii) the nonlinear absorption process does not require any photosensitivity of the material, facilitating fabrication in glasses, crystals, polymers, and practically any optical material. Although preprocessing of the materials to be inscribed is not necessary, it can be helpful. Hydrogen loading, for example, can enhance the sensitivity to inscription of fiber Bragg gratings [8, 9].

Different categories of index change have been defined in the literature, mostly with respect to grating fabrication. Type I index changes happen for pulse energies

IntechOpen

close to the nonlinear ionization threshold (10^{13} W/cm^2) and cause an accumulative change in the refractive index of the order of 10^{-3} (in silica glass). The change in the refractive index is isotropic and is mostly attributed to localized material melting and rapid resolidification [10, 11], although other explanations (such as color center formations) are also considered [12]. This type of index change is most useful for the fabrication of waveguides [13], couplers [14], and FBGs [15].

Type II interaction happens at intensities beyond the damage threshold, which can lead to the formation of voids [16]. Voids are submicron features, micro-explosions in matter, or air bubbles, with larger refractive index contrast compared to their surroundings. They are achieved by extremely tight focusing with power densities of the order of 10^{15} W/cm^2. Voids attract interest mainly due to their potential as permanent highly dense 3D optical data storage materials. In such schemes, each void represents a bit, which can be read with transmitted or scattered light. It was found that voids can also be seized, moved, and merged by femtosecond laser radiation [17]. Type II FBGs, also termed "damage" gratings, have been shown to withstand higher temperatures and can be used as harsh environment sensors [18].

On applying intensities between the above regimes, an anisotropic, polarization-dependent, index change is induced, and the glass material becomes birefringent [19, 20]. The magnitude of the reported index change is the same as for type I changes, but it is not isotropic. The intensity boundaries for this interaction are not well defined, as they depend on the laser source, the focusing lens, and the material itself. The anisotropy of the refractive index change is believed to originate from the nanogratings observed inside the focal volume. The planes of these gratings are perpendicular to the light polarization and behave as negative uniaxial crystals [21–24].

In the following we will focus on fabrication of FBGs using femtosecond laser. Section 2 briefly describes methods of fabrications using femtosecond laser and references to a more detailed work on the subject. Section 3 introduces the main concept of this chapter—transient fiber Bragg gratings for optical switching. The theory of transient grating is outlined, and an overview of various works on the subject is described. Section 4 provides experimental results achieved by our group on generation and characterization of transient FBGs. Finally, we summarize and discuss possible future research direction of transient Bragg grating switching.

2. Femtosecond inscription of fiber Bragg gratings

FBG fabricated with femtosecond laser was first demonstrated by the point-by-point (PbP) method [25, 26]. In this method, the beam is tightly focused into the fiber core to a spot size radius smaller than half of the desired grating period. To achieve this, a microscope objective with a high numerical aperture must be used, as well as pulse energies just above the inscription threshold. The induced index change happens on the pulse peak intensity only, which can be smaller than the diffraction limit of the focusing objective lens. To fabricate the grating, the fiber is aligned and translated in the focal plane at constant velocity. The scan velocity matches the grating period to the laser pulse rate, so that each pulse inscribes a single grating "plane."

The PbP method requires tight control on all-optical and mechanical parameters of the system. The optical system must be carefully aligned to avoid aberrations and achieve the smallest spot size. The pulse width and energy should be controlled as well, since they affect the actual spot size. For this reason, most PbP systems use 800 nm femtosecond laser, rather than its harmonics, to avoid dispersions [27].

From a mechanical perspective, the fiber core must be maintained in the focal plane through the entire fabrication process. This requires high-end air-bearing translation stages. An extension of the PbP method to reduce the mechanical complexity is the line-by-line method, in which the beam is scanned across the fiber axis and forms a rectangular "snake" pattern [28], or plane-by-plane method in which the beam is focused to an elliptic sheet, creating 2D index change [29].

The PbP method offers the highest flexibility in grating fabrication. Uniform gratings, phase-shifted gratings [30], apodization [31], and more [32] have been demonstrated. The tight focusing condition also enables inscription through the fiber jacket without damage [33]. The same inscription system can be used for the fabrication of waveguides and long-period gratings as well [34–36]. Gratings fabricated by this method have been shown to have superior thermal properties [37] than UV gratings and better performance as fiber laser mirrors [38–40].

In 2003, Mihailov et al. demonstrated the fabrication of FBGs with a femtosecond laser and a phase mask [41, 42]. The optical configuration is similar to its UV counterpart. The beam is focused on the fiber core using a cylindrical lens and through a phase mask. The mask period defines the Bragg grating period. In the phase mask configuration, the grating is inscribed as a whole rather than plane by plane. It is robust, repeatable, and typically stationary. As the period is defined by the phase mask, relatively long-focus lenses can be used, which greatly eases alignment and makes this configuration suitable for large core fibers as well. With this technique, grating inscription has been demonstrated in various types of fibers [43–46]. The main drawback of this configuration is the lack of flexibility, as the period is predetermined by the phase mask. Nevertheless, it is possible to tune the Bragg wavelength by introducing defocusing and other aberrations into the inscribing beam. Shifts of more than 300 nm, as well as chirp gratings, have been demonstrated with this method [47, 48]. Inscription through the coating is also feasible in this method with "of the shelf" high-NA cylinder lenses [49–51].

Both methods have been used for fabrication of fiber Bragg grating with superior properties than grating fabrication with UV sources. Femtosecond laser can be used to fabricate gratings in any type of fibers and can withstand higher temperature than UV gratings. The most notable feature is the ability to inscribe grating through the fiber coating, thus maintaining its mechanical strength, and avoid handling issues such as stripping, cleaning, and recoating [52–54].

3. Transient fiber Bragg grating optical switching

All-optical switching has been investigated for a long time by the optical community, in particular for optical communication applications. If successful, it will dramatically increase the throughput in optical links and will enable data switching at speeds and rates far beyond the capabilities of current electronic devices.

Recently, there has been a growing interest in FBGs for optical switching applications. Several works reported implementations of an optical switch by tuning a pre-inscribed grating by means of heat, stress, and other relatively slow processes [55–58]. These methods are based on permanent FBGs, in which any change in the refractive index (heat, cross-phase modulation) or period (induced stress) will shift the grating resonance from the signal wavelength. Such switching mechanisms have several drawbacks due to the inherent physical properties of their operation, which limits their applicability and performance. In the wider context, there have been several reports on the switching of various photonic crystal structures, both for fundamental and for applicative purposes (see, e.g., [59, 60] for some recent reviews).

Transient Bragg gratings (TBGs) can overcome these limitations. These are Bragg gratings of finite duration. In the case of femtosecond gratings in materials, they are expected to be formed at intensities below the threshold for permanent index modification and to exist for the inscribing pulse duration only. Transient gratings in fibers or waveguides are expected to act as a fast switch or modulator by implementing a Bragg mirror with (ultra-) fast decay time.

Several mechanisms are available to implement transient Bragg gratings: the optical Kerr effect, free-carrier recombination in semiconductor materials, and diffusion of thermal gratings. The different mechanisms differ from one to another by the rise and decay time of the switch and by the extinction ratio, i.e., the contrast between on and off states. Such transient gratings can be turned on/off by modulating the illumination beam.

The Kerr effect describes the refractive index change in the presence of high intensities, such as those that are available from high-power femtosecond lasers [61]. The refractive index changes by an amount of $n_2 I$, where n_2 is the material nonlinear index and I is the intensity. The response of the material is instantaneous. For silica fibers, $n_2 \sim 3 \cdot 10^{-16}$ cm^2/W; thus, for an intensity of $I = 10^{11}$ W/cm^2, the refractive index change is of the order of 10^{-5}. Stronger index change is feasible for materials with higher Kerr nonlinearity, such as Chalcogenide or Bismuth fibers [43, 62]. The Kerr grating has a periodic pattern, with the index modulation as described above. A Kerr grating switch is expected to be weak yet with a femtosecond time scale response. Several publications reported on transient Kerr gratings in gas for the purpose of spectroscopy and in bulk semiconductors for studying free-carrier recombination rates [63]. An optical grating based on the nonlinear Kerr effect has been used in the past for parametric wavelength conversion [64] and for chemical spectroscopy [65]. An optical switch based on an optical Kerr grating has only been investigated numerically until now [66–68].

Free-carriers in semiconductor materials are formed upon pulse irradiation followed by excessive charge concentrations. In this case, the refractive index changes due to different charge densities are much higher than due to the Kerr effect. The transient index change of the semiconductor is described by the Drude model of excited free-carriers [69–71] reaching values as high as $\delta n/n \sim 10^{-1}$ [59, 60]. Unlike the Kerr effect, which is instantaneous, the FC excitation is "turned on" fast but typically persists for a time scale of several tens of picosecond to several ns depending on the recombination rate of the generated electron-hole pairs and the diffusion length [63, 72–74]. An optical switch-based free-carrier transient Bragg grating is expected to have better contrast and stronger reflection but on a much shorter switching time. As the reflection is very sensitive to the grating period (typically 1 µm), extremely small diffusion is sufficient to wash out the grating and its reflection. Sivan et al. showed theoretically that when exciting a transient grating based on FC, the turn-off times are very fast (<ps) due to diffusion of the excited FCs that erases the grating structure [75]. This is a key point that will allow for switching times several orders of magnitude faster than in bulk FC switching configurations, thus potentially revolutionizing switching technology. Such an optical switch is expected to have a better extinction ratio than a Kerr grating and slower (picosecond) time scale switching.

The same principle can be applied for transient thermal gratings, in which the index changes as a response to localized heat or increased temperature and the diffusion length is determined by the material properties. Transient thermal gratings are used as a method for measuring the diffusion coefficient of materials and were implemented in opaque materials with linear absorption at the laser wavelength [76–78]. Optical materials are mostly transparent to NIR femtosecond lasers

(800 nm); therefore, a transient thermal grating may only be realized through *nonlinear* absorption.

Another mechanism is to form dynamic population gratings in active fibers. This was implemented via counter propagating waves and resulted in millisecond time responses [79].

TBGs of a few centimeter lengths were implemented using 193 nm, nanosecond, excimer laser pulses, and a phase mask in phosphosilicate fibers without hydrogen loading. In passive fibers, extremely slow reflection of tens of seconds' duration was demonstrated [80], while in active fibers, the grating was based on population inversion, and the time response was estimated to be milliseconds long [81]. In both cases the expected rise time of such switching mechanism cannot be shorter than the ns pulse length.

Transient grating-based switching was suggested numerically by coupling light from the fundamental mode to high-order modes [82]. Nanosecond switching was implemented using the Kerr effect with a highly nonlinear polymer layer deposited close to the core of a polished fiber [83]. It was suggested, theoretically, that a TBG would result in an ultrafast switching response [84]. Thermal phenomena are typically associated with relatively long (microsecond) time scales. Recently, it was suggested, theoretically, that picosecond scale switching is achievable with thermal gratings, using metal nanoparticles in waveguides [85]. Nanosecond switching of a permanent FBG was demonstrated by introducing electrodes into a special two-hole fiber [86]; however, this device suffers from nanosecond rise time and a millisecond time scale to return to its original state.

TBGs essentially enable pulse extraction from CW source. This can lead to several photonic applications such as all-optical switching and modulator at any wavelength, all-fiber Q-switching mechanism, and sub-ns pulse sources.

In the following, we will shortly describe the theory of transient Bragg grating. A detailed derivation of the suitable coupled mode equations, and their numerical solution can be found in literature. Here, we begin our discussion from the coupled mode equations and limit the discussion to specific case where analytical solution is possible to gain physical insight. Next, we will describe our group experimental work on transient Bragg gratings in silica fibers. We will show the dynamic of permanent grating switching and describe an immunization technique that enable, for the first time to our knowledge, thermal grating-based nonlinear absorption.

The theory of transient Bragg gratings is fully developed and described in the literature [87, 88] starting from the wave equation. Here, we provide a short description of the theory starting from its derived coupled mode equation for transient grating. The analysis begins from the well-known coupled mode equations for forward and backward propagating waves in grating media, adapted to the case of transient grating:

$$\frac{d}{dt}A_f + v_g\frac{d}{dz}A_f + 2\,iv_g\kappa q(z)m(t)A_f = iCv_g\kappa q(z)m(t)e^{(-2i\delta\kappa z)}A_b \qquad (1)$$

$$\frac{d}{dt}A_b - v_g\frac{d}{dz}A_b + 2\,iv_g\kappa q(z)m(t)A_b = -iCv_g\kappa q(z)m(t)\exp^{(2i\delta\kappa z)}A_f \qquad (2)$$

Here, A_f and A_b represent the envelopes of the forward and backward pulses, respectively, v_g is the group velocity, $q(z)$ is the spatial shape of the inscribed Bragg grating, and $m(t)$ is its temporal profile. C is the grating contrast, and $\delta k = \delta\omega/v_g$ is the detuning of the incident pulse from the center of the spectral gap. The forward-backward mode coupling coefficient, $\kappa = k_0 n_0 \Delta n/4n_{neff}$, is a product of the free space wavevector k_0, the waveguide material refractive index n_0, and its maximal modulation amplitude Δn, divided by the effective mode index n_{neff}. These equations are similar to those obtained in [87]; however, they account for non-zero mean index modulations, absorption, imperfect grating contrast, and nonuniform pumping (via $q(z)$).

An exact solution of Eqs. (1) and (2) is possible only numerically. However, if one assumes uniform pump spot ($q = 1$) and ignores the spatial derivatives (justified for short modulations during which the pulse is nearly stationary), Eqs. (1) and (2) can be solved analytically—this yields the well-known Rabi solution (see, e.g., [87, 89]). This was shown to give a reasonable accuracy in measurements with spin waves [87], at least in this limit, and to lead to envelope reversal [90–92].

Alternatively, Eqs. (1) and (2) can also be solved analytically in the low conversion efficiency limit, without neglecting the spatial derivatives. In this case, the efficiency of the backward pulse generation is given approximately by a convolution of the incoming pulse with $m_{eff}(t)$, where $M_{eff}(t)$ is an *effective* modulation, and the forward wave is (nearly) monochromatic (e.g., for a CW or nanosecond source $—A_f(t) \rightarrow 1$):

$$A_b(z,t) = iCv_g\kappa q(z)m(t)e^{\{2i\delta kz\}}M_{eff}(t) \tag{3}$$

$M_{eff}(t) = q(z) * m(t)$ is the convolution of the (transverse) spatial and temporal profiles of the pump pulse. These equations can be solved analytically assuming symmetric Gaussian shape for the pump pulse: $q(z) = e^{(-z/L_g)^2}$ and $m(t) = e^{(-t/T_{mod})^2}$.

The complex analytical solution for the backward reflected pulse depends on two time scales: (i) the modulation time T_{mod}, which, for a Kerr grating, is the pump pulse duration, and (ii) the grating pass time $T_{pass} = L/v_g$, where L is the grating length and has the form:

$$|A_b(t,z)| \sim \sqrt{\pi}C\frac{n_0}{4n_{eff}^2}\Delta n\omega_0\sqrt{\frac{T_{pass}^2 T_{mod}^2}{T_{pass}^2 + T_{mod}^2}}e^{\left\{-\frac{(z+v_gt)^2}{v_g^2\left(T_{pass}^2+T_{mod}^2\right)}\right\}} \tag{4}$$

Significant physical insight is achieved under the assumption that $T_{mod} \ll T_{pass}$. The solution for the reflected wave is then

$$|A_b(t,z)| \sim \sqrt{\pi}C\frac{n_0}{4n_{eff}^2}\Delta n\omega_0 T_{mod}e^{\left\{-\frac{(z+v_gt)^2}{v_g^2 T_{pass}^2}\right\}} \tag{5}$$

This reveals unique spatial-temporal dependency. The reflected wave has the temporal duration of the longer time scale, and the power is scaled as the shorter time scale. This occurs because the reflections occur from within the grating rather than outside of it. Note that the grating length only influences the temporal duration and not the power efficiency. The reflected efficiency can be approximated to be $\sim \left(\frac{\pi^{\frac{3}{2}}}{2\lambda_0 n_0}\Delta n c_0 T_p\right)^2$.

For a 10^{-4} index change and a 50-fs pump duration with a 1500-nm signal wavelength, we get an efficiency of $\sim 4 \cdot 10^{-6}$. Since $v_g T_p \sim 10\ \mu m$, then the minimal length of the grating for this limit to hold is about 0.1 mm. The backward pulse is then at least 500 fs long.

In the opposite case of a very short grating, $T_{mod} \gg T_{pass}$, as is feasible in semiconductors, the reflected pulse power is

$$|A_b(t,z)| \sim \sqrt{\pi}C \frac{n_0}{4n_{eff}^2} \Delta n \omega_0 T_{pass} e^{\left\{-\frac{(z+v_g t)^2}{v_g^2 T_{mod}^2}\right\}} \tag{6}$$

Thus, the reflected wave duration and spectrum follow that of the pump pulse.

The above approximations are valid for low reflection efficiency, i.e., un-depleted pump. Transient Kerr grating is expected to have a very low efficiency, and an order of magnitude difference between T_{pass} and T_{mod} is expected to correspond to the above solutions.

In silica fibers, there is also the possibility for thermal gratings. In this case the index modulation time varies on the microsecond and nanosecond time scales, which is considerably longer than the passage time for a typical 5-mm grating (~25 ps). The expected reflectivity should behave as in the first case above with the exception of non-Gaussian response.

The results indicate that with the Kerr mechanism high reflection efficiencies are feasible for Chalcogenide fibers and semiconductor waveguides; however, silica fibers are more challenging. Furthermore, as the reflection from transient Bragg gratings is dependent on the pumping configuration, e.g., grating length and pump pulse duration, and it is possible to control the signal modulation. In the theory, generation pulses on time scale such as tens of ps, currently not available from fiber lasers, are possible. Other interesting applications such as in-fiber Q-switching are also feasible.

In the next sections, we will describe experimental results achieved by our group on the subject of transient Bragg gratings in standard silica fibers for switching and modulation applications. We will describe methods to generate them and their results.

4. Experimental results of femtosecond transient FBGs

4.1 Experimental setup

The experimental setup is standard for FBG inscription with the phase mask technique and is shown schematically below (**Figure 1**). A femtosecond laser (800 nm, 35 fs, 1 KHz) is focused on the fiber core, through a phase mask. The mask period is 2.14 µm, suitable for second-order Bragg gratings at 1550 nm. The fiber to be inscribed is connected to a probe signal source and an Optical Spectrum Analyzer (OSA) to monitor the FBG spectrum or to a fast photodiode (Thorlabs DETO8CFC) to monitor the dynamic effects. The signal source can be a broadband ASE source when characterize permanent FBG inscription or an amplified DFB laser when observing transient, dynamic effect. The probe laser mostly operated in CW mode providing 1 W output power and was operated in pulse mode for Kerr grating experiments.

4.2 Transient Kerr grating

We tried to observe a transient Kerr grating with pulse energies below the inscription threshold in standard SMFs. In these experiments, we monitor the reflection from the grating with a photodiode. We found the permanent inscription threshold to be 160 µJ; thus, our pulse energy is limited below this value. For 100 µJ pulse energy, we expect grating index modulation of $8 \cdot 10^{-7}$, which will exist for 35 fs only. The expected reflection from such a grating is extremely weak; the

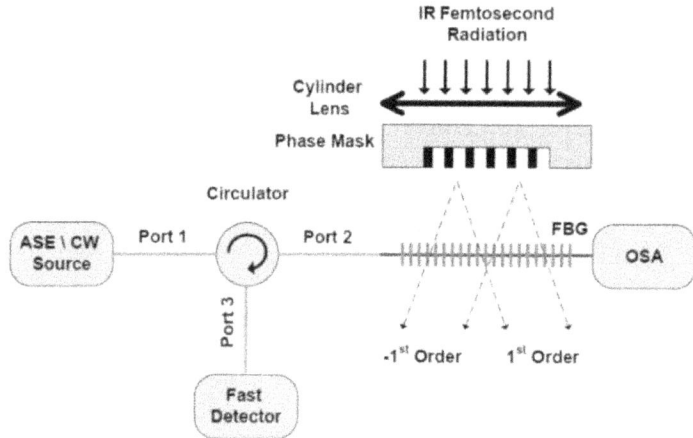

Figure 1.
Schematic of the optical setup.

coupling coefficient, calculated according to the theory outlined in Section 3 is $\kappa \sim 8 \cdot 10^{-7}$ $\left[\frac{1}{\mu m}\right]$, four orders of magnitude lower than typical permanent gratings. Therefore, we drive our probe laser with 50 ns pulses at a 20 KHz pulse rate. In this mode, the laser outputs 1 KW peak power, tuned to the Bragg wavelength. The reflected efficiency expected for such index modulation is $\sim 10^{-5}$.

Unfortunately, we could not detect any Kerr grating reflections with our detector or with a lock-in amplifier. Furthermore, we noticed an increase in the detector DC level, and a photodiode was able to measure a weak (nW) but slowly growing reflected power signal indicating permanent inscription. We repeated the experiment with 50 µJ pulse energy to find again permanent inscription.

The reflected power was extremely weak and could not be detected with a standard ASE source. The permanent inscription may be the result of tunneling ionization rather than multiphoton ionization, which is a much slower process that is expected for relative low intensities by the Keldysh theory [93]. Further investigation is required in order to produce Kerr grating, most likely in a different material with higher nonlinearity. In the following we will present different methods to observe transient gratings based on thermal effects in silica fibers.

4.3 Permanent grating switching

In this configuration, we first fabricate a high-quality (>25 dB) grating and observed light transmitted through it, i.e., we measure the transmission loss of the grating rather than its reflections. To modulate the grating, we block half of the beam and illuminate only half of the permanent grating through the phase mask. Due to the induced heat of each pulse, the refractive index is elevated, causing half the grating to shift to a higher Bragg wavelength, leaving the other half intact. This opens a transmission gap in the grating, allowing a signal, at wavelength matching the grating Bragg wavelength to be detected by the photodiode. Essentially, we temporarily transform a uniform grating into a phase-shifted grating. **Figure 2** shows the time profile of the transmitted signal through the shifted grating.

The switching mechanism is based on induced heat, as if the gratings were placed on a temperature-controlled controller. However, here the switching is done with an ultrafast laser that provides ultrafast rise time. As can be seen in **Figure 2**, the switching time here is about 8 µs, which makes it suitable for Q-switching

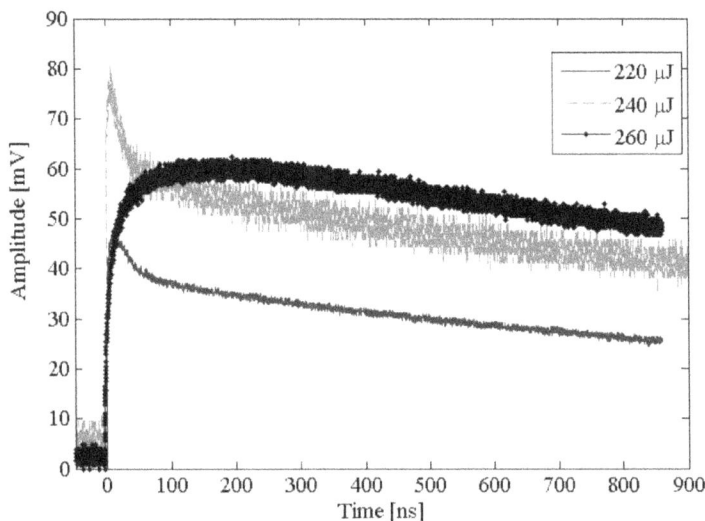

Figure 2.
Pulse measured out of a modulated permanent grating as a function of pulse energy.

applications. We can reduce the switching time to less than 2 µs at the cost of extinction ratio by setting the signal wavelength slightly away from the grating resonance.

While this is a very slow modulation time, it shows the natural response of the induced grating by femtosecond laser pulse at different pulse energies above the inscription threshold. Note that two different regimes are noticeable: for low pulse energies, a fast decay of the signal followed by a long µs tail. When increasing the pulse energy, the fast decay disappears. This indicates the formation of permanent index modification. We will show that the long µs tail can be cut off by performing immunization.

4.4 Immunization to femtosecond inscription

The ability of femtosecond lasers to modify the material refractive index of practically any optical material is a keystone in photonic device fabrication. However, when one wants to observe transient grating effects, it is a drawback, as it limits the applied pulse energy to energies below the multiphoton ionization threshold and permanent index change. It is known that the modified index has a limit, i.e., it can only grow to the order of 10^{-3} (positive change in silica fibers) when it reaches saturation. Reflection from a transient grating, however, depends on both the index modulation and the grating period.

We have found that femtosecond photo pre-treatment can immunize a fiber up to a certain illumination intensity [94]. In fact, the reported multiphoton ionization and inscription threshold ($\sim 10^{13}$ W/cm^2) can be raised so that permanent Bragg gratings first appear at a higher pulse energies.

Fiber immunizing can help avoid permanent index change and observation of transient FBG effects. After immunization, the fiber transient index change effects, such as heat or Kerr, are expected to be observed more clearly as the permanent index change is saturated and its effects are suppressed.

In order to immunize the fiber against femtosecond inscription, we remove the phase mask and inscribe it with a focal line pattern. The pulse energy in this

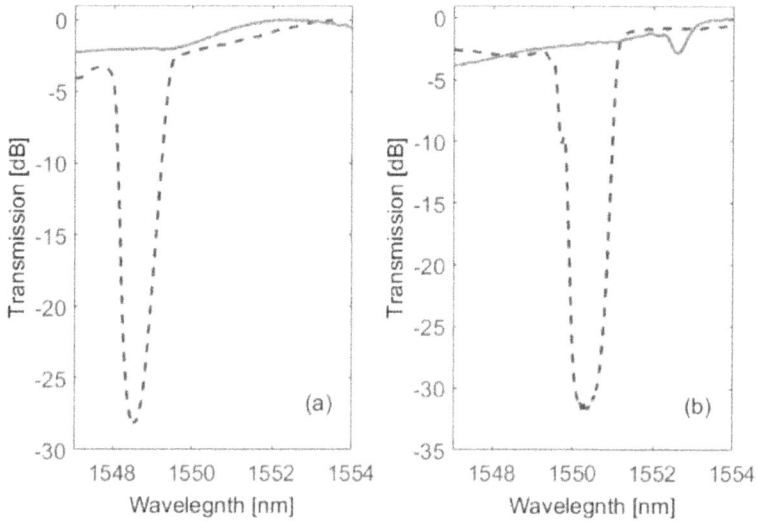

Figure 3.
Transmission spectra of the inscribed permanent FBGs in untreated and treated fibers: untreated fiber, dashed blue lines; treated fibers, solid red lines. (a) 160 μJ pulse energy. (b) 180 μJ pulse energy.

pre-treatment was chosen to be slightly more than twice the average energy for FBG inscription, so that the peak intensity of the pre-treatment would be slightly higher than that of the FBG inscription.

Figure 3 shows permanent gratings inscribed on a fresh fiber compared to a treated fibers. Before any pre-treatment, gratings with −25 and − 30 dB transmission dips were inscribed, at pulse energies of 160 μJ (**Figure 3a**) and 180 μJ (**Figure 3b**), respectively. As is evident, when pre-treatment of the fibers is done at slightly more than double of the pulse energy, the results is a complete immunity for inscription at 160 μJ and in only −2 dB transmission loss at 180 μJ. For the latter, a 2.5-nm wavelength shift is observed, corresponding to an increase of $\sim 2 \cdot 10^{-3}$ of the average refractive index due to the pre-treatment. Thus, our treatment greatly reduces the ability to inscribe gratings. Pre-treatment of the fiber causes complete immunity or limited grating buildup at a considerably lower rate.

4.5 Fast switching with transient thermal grating

Fiber immunization extremely limits the grating buildup. We characterized the transient grating reflections of an immunized fiber [95]. After completing the photo-treatment process on a standard SMF at a pulse energy of 1 mJ, the pump laser pulse rate was lowered to 2 Hz in order to reduce the average thermal effects, and reflected pulses were measured with our detector. **Figure 4a** shows the averaged time trace (100 pulses) of the reflected pulse for different femtosecond illumination pulse energies (all below half of the immunization pulse energy).

The reflected pulses (**Figure 4a**) have a very fast rise time followed by nanosecond decay. This is three orders of magnitude improvement compared to transient grating based on UV laser reported in Ref. [80, 81]. The observed transient increase in the reflectivity can mainly be attributed to local heating of the silica, due to *nonlinear* absorption, corresponding to a local increase in the refractive index and is followed by thermal diffusion that washes out the grating. The decay time is typical for thermal diffusion at these sizes and distances.

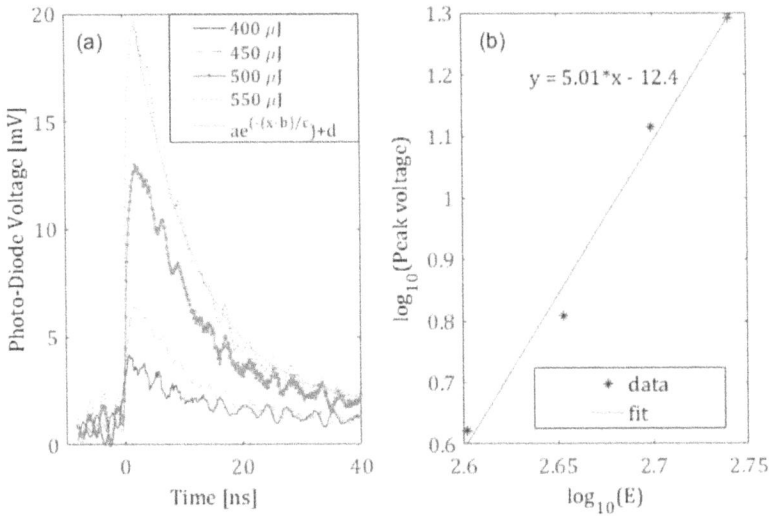

Figure 4.
(a) Measured average time trace of the reflected pulses for different femtosecond illumination pulse energies. The inscribing pulse rate was 2 Hz. An exponential fit with parameters, a = 17.87, b = 2.06, c = 10.5, and d = 1.7, is shown for the highest reflectivity, indicating a thermal diffusion time of ~10 ns. (b) The peak amplitude of the measured reflected signal as a function of pulse energy. The linear fit indicates a nonlinear relation between the pulse energy (intensity) and reflectivity.

The measured rise time is 2 ns, but we believe the actual rise time is significantly shorter, since our measurement was limited by our detection system (about the same rise time was also measured for our 35 fs laser source). For a 550 μJ illumination pulse, the reflected pulse duration is approximately 14.3 ns (measured between 1 and e points). An exponential fit is shown (light blue) in **Figure 4a**, with a time constant of ~10 ns. For a 550-μJ illumination pulse, the measured reflected peak power is 0.38 mW, corresponding to a peak power reflectivity of 0.0435%.

Assuming an effective grating length of 5 mm, and applying the theory for spatial-temporal gratings, we estimate an index change of $\Delta n = 2.3 \cdot 10^{-6}$. This is three orders of magnitude less than reported in the literature for permanent femtosecond inscribed gratings. However, it should be noted that tens of thousands of pulses are used to achieve the reported Δn for permanent inscription. The thermal grating, and the refractive index increase, decay time depends on the diffusion coefficient of the fiber and the grating period. In this case, the decay time expected to be 34 ns [96]. The reflected power from a temporal grating is proportional to $(\Delta n)^2$; thus, we expect from theory (Eq. (6)) a reflected signal decay time of 17 ns. This is in good agreement with the experimental results, where differences may arise due to the presence of germanium in the fiber core.

Figure 4(b) shows the peak reflectivity as a function of applied pulse energy. A small increase in the inscribing pulse energy results in a higher induced transient refractive index change, leading to a significantly stronger reflected pulse. As suggested by the linear fit, the peak reflectivity has, indeed, a nonlinear growth that corresponds to I^5, which is a good indication that the grating is, indeed, based on multiphoton absorption.

With respect to **Figure 2**, and time scale reported with femtosecond induced index change [4], the immunization technique allows us to remove transient effects associated with material resolidification and access the previous phase of femtosecond laser-matter interaction.

We also note here that thermal grating diffusion time is highly dependent on grating period. The diffusion time is opposite to the square of the grating period; thus, working with first-order grating can reduce the time scale by a factor of four. The applicability to transient thermal grating for higher pulse rate and life time of such device is elaborated elsewhere [95].

5. Conclusions

Transient fiber Bragg gratings has a great potential for all-optical, all-fiber, fast optical switching. Many challenges have yet remained to be investigated in this field, mostly improving the efficiency of the grating for practical applications, methods to generate them, and life time of such devices.

Achieving Kerr gratings for ultrafast switching is challenging in silica fibers since the effect is much weaker than inscription of permanent gratings. However, it may be feasible in highly nonlinear fibers or waveguide materials. Semiconductor waveguides and materials are promising for both Kerr-based transient gratings and free-carrier-based gratings. Furthermore, it should be possible to implement with low-power, low-cost diodes rather than high-power femtosecond lasers—at the cost of slower rise time.

The immunization technique presented here can be used to implement transient thermal gratings in transparent materials and may serve as a diagnostic tool for dielectric materials with different compositions and doping. Furthermore, the diffusion of transient thermal gratings is highly dependent on the grating period; thus, many time scales and wavelengths are accessible by simply choosing a suitable mask and illuminating wavelength.

Several applications may rise from transient gratings in fibers and remain to be demonstrated: fiber laser Q-switching and modulation, generation of sub-ns pulses—a regime not accessible with Q-switching or mode-locking technique, diagnostic tools, and more.

Acknowledgements

We would like to thank Dr. Yonatan Sivan from the Electro-Optics Unit in Ben-Gurion University of the Negev, for his contribution and help in understanding transient grating theory.

Author details

Avishay Shamir[1,2]*, Aviran Halstuch[1] and Amiel A. Ishaaya[1]

1 Department of Electrical and Computer Engineering, Ben-Gurion University of the Negev, Beer-Sheva, Israel

2 Israel Center of Advanced Photonics (sICAP), Yavne, Israel

*Address all correspondence to: avishay.shamir@gmail.com

IntechOpen

© 2019 The Author(s). Licensee IntechOpen. This chapter is distributed under the terms of the Creative Commons Attribution License (http://creativecommons.org/licenses/by/3.0), which permits unrestricted use, distribution, and reproduction in any medium, provided the original work is properly cited. (cc) BY

References

[1] Banks PS, Stuart BC, Komashko AM, Feit MD, Rubenchik AM, Perry MD. Femtosecond laser materials processing. In: Symposium on High-Power Lasers and Applications. International Society for Optics and Photonics. 2000. pp. 14-21

[2] Smith G, Kalli K, Sugden K. Advances in femtosecond micromachining and inscription of micro and nano photonic devices. Frontiers in Guided Wave Optics and Optoelectronics, Bishnu Pal, IntechOpen, DOI: 10.5772/39542. Available from: https://www.intechopen.com/books/frontiers-in-guided-wave-optics-and-optoelectronics/advances-in-femtosecond-micromachining-and-inscription-of-micro-and-nano-photonic-devices; 2010

[3] Homoelle D, Wielandy S, Gaeta AL, Borrelli NF, Smith C. Infrared photosensitivity in silica glasses exposed to femtosecond laser pulses. Optics Letters. 1999;**24**(18):1311-1313

[4] Gattass RR, Mazur E. Femtosecond laser micromachining in transparent materials. Nature Photonics. 2008;**2**(4): 219-225

[5] Thomas J, Voigtlaender C, Becker RG, Richter D, Tuennermann A, Nolte S. Femtosecond pulse written fiber gratings: A new avenue to integrated fiber technology. Laser & Photonics Reviews. 2012;**6**(6):709-723

[6] Della Valle G, Osellame R, Laporta P. Micromachining of photonic devices by femtosecond laser pulses. Journal of Optics A: Pure and Applied Optics. 2008;**11**(1):013001

[7] Beresna M, Gecevičius M, Kazansky PG. Ultrafast laser direct writing and nanostructuring in transparent materials. Advances in Optics and Photonics. 2014;**6**(3):293-339

[8] Kryukova PG, Larionova YV, Rybaltovskiia AA, Zagorul'koa KA, Dragomirb A, Nikogosyanb DN, et al. Long-period fibre grating fabrication with femtosecond pulse radiation at different wavelengths. Microelectronic Engineering. 2003;**69**(2):248

[9] Bernier M, Sheng Y, Vallée R. Ultrabroadband fiber Bragg gratings written with a highly chirped phase mask and infrared femtosecond pulses. Optics Express. 2009;**17**:3285-3290

[10] Chan JW, Huser T, Risbud S, Krol DM. Structural changes in fused silica after exposure to focused femtosecond laser pulses. Optics Letters. 2001; **26**(21):1726-1728

[11] Bellouard Y, Colomb T, Depeursinge C, Dugan M, Said AA, Bado P. Nanoindentation and birefringence measurements on fused silica specimen exposed to low-energy femtosecond pulses. Optics Express. 2006;**14**(18): 8360-8366

[12] Zoubir A, Richardson M, Canioni L, Brocas A, Sarger L. Optical properties of infrared femtosecond laser-modified fused silica and application to waveguide fabrication. Journal of the Optical Society of America: B. 2005; **22**(10):2138-2143

[13] Miura K, Qiu J, Inouye H, Mitsuyu T, Hirao K. Photowritten optical waveguides in various glasses with ultrashort pulse laser. Applied Physics Letters. 1997;**71**(23):3329-3331

[14] Streltsov AM, Borrelli NF. Fabrication and analysis of a directional coupler written in glass by nanojoule femtosecond laser pulses. Optics Letters. 2001;**26**(1):42-43

[15] Mihailov SJ, Grobnic D, Smelser CW, Lu P, Walker RB, Ding H. Induced

Bragg gratings in optical fibers and waveguides using an ultrafast infrared laser and a phase mask. Laser Chemistry. 2008;**2008**. Article ID: 416251, 20 pages. https://doi.org/10.1155/2008/416251

[16] Glezer EN, Mazur E. Ultrafast-laser driven micro-explosions in transparent materials. Applied Physics Letters. 1997;**71**(7):882-884

[17] Watanabe W, Toma T, Yamada K, Nishii J, Hayashi KI, Itoh K. Optical seizing and merging of voids in silica glass with infrared femtosecond laser pulses. Optics Letters. 2000;**25**(22): 1669-1671

[18] Smelser CW, Mihailov SJ, Grobnic D. Formation of type I-IR and type II-IR gratings with an ultrafast IR laser and a phase mask. Optics Express. 2005; **13**(14):5377-5386

[19] Sudrie L, Franco M, Prade B, Mysyrowicz A. Study of damage in fused silica induced by ultra-short IR laser pulses. Optics Communications. 2001;**191**(3):333-339

[20] Bricchi E, Klappauf BG, Kazansky PG. Form birefringence and negative index change created by femtosecond direct writing in transparent materials. Optics Letters. 2004;**29**(1):119-121

[21] Bhardwaj VR, Simova E, Rajeev PP, Hnatovsky C, Taylor RS, Rayner DM, et al. Optically produced arrays of planar nanostructures inside fused silica. Physical Review Letters. 2006; **96**(5):057404

[22] Beresna M, Kazansky PG. Polarization diffraction grating produced by femtosecond laser nanostructuring in glass. Optics Letters. 2010;**35**(10):1662-1664

[23] Tang Y, Yang J, Zhao B, Wang M, Zhu X. Control of periodic ripples growth on metals by femtosecond laser

ellipticity. Optics Express. 2012;**20**(23): 25826-25833

[24] Hnatovsky C, Shvedov V, Krolikowski W, Rode A. Revealing local field structure of focused ultrashort pulses. Physical Review Letters. 2011; **106**(12):123901

[25] Martinez A, Dubov M, Khrushchev I, Bennion I. Direct writing of fibre Bragg gratings by femtosecond laser. Electronics Letters. 2004;**40**(19):1

[26] Geernaert T, Kalli K, Koutsides C, Komodromos M, Nasilowski T, Urbanczyk W, et al. Point-by-point fiber Bragg grating inscription in free-standing step-index and photonic crystal fibers using near-IR femtosecond laser. Optics Letters. 2010;**35**(10): 1647-1649

[27] Dubov M, Bennion I, Nikogosyan DN, Bolger P, Zayats AV. Point-by-point inscription of 250 nm period structure in bulk fused silica by tightly focused femtosecond UV pulses. Journal of Optics A: Pure and Applied Optics. 2008;**10**(2):025305

[28] Zhou K, Dubov M, Mou C, Zhang L, Mezentsev VK, Bennion I. Line-by-line fiber Bragg grating made by femtosecond laser. IEEE Photonics Technology Letters. 2010;**22**(16): 1190-1192

[29] Theodosiou A, Lacraz A, Polis M, Kalli K, Tsangari M, Stassis A, et al. Modified fs-laser inscribed FBG array for rapid mode shape capture of free-free vibrating beams. IEEE Photonics Technology Letters. 2016;**28**(14): 1509-1512

[30] Burgmeier J, Waltermann C, Flachenecker G, Schade W. Point-by-point inscription of phase-shifted fiber Bragg gratings with electro-optic amplitude modulated femtosecond laser pulses. Optics Letters. 2014;**39**(3): 540-543

[31] Williams RJ, Voigtländer C, Marshall GD, Tünnermann A, Nolte S, Steel MJ, et al. Point-by-point inscription of apodized fiber Bragg gratings. Optics Letters. 2011;**36**(15): 2988-2990

[32] Marshall GD, Williams RJ, Jovanovic N, Steel MJ, Withford MJ. Point-by-point written fiber-Bragg gratings and their application in complex grating designs. Optics Express. 2010;**18**(19):19844-19859

[33] Martinez A, Khrushchev IY, Bennion I. Direct inscription of Bragg gratings in coated fibers by an infrared femtosecond laser. Optics Letters. 2006; **31**(11):1603-1605

[34] Zhang H, Eaton SM, Li J, Herman PR. Femtosecond laser direct writing of multiwavelength Bragg grating waveguides in glass. Optics Letters. 2006;**31**(23):3495-3497

[35] Li B, Jiang L, Wang S, Tsai HL, Xiao H. Femtosecond laser fabrication of long period fiber gratings and applications in refractive index sensing. Optics & Laser Technology. 2011;**43**(8): 1420-1423

[36] Fujii T, Fukuda T, Ishikawa S, Ishii Y, Sakuma K, Hosoya H. Characteristics improvement of long-period fiber gratings fabricated by femtosecond laser pulses using novel positioning technique. In: Optical Fiber Communication Conference. Optical Society of America; 2004. p. ThC6

[37] Martinez A, Khrushchev IY, Bennion I. Thermal properties of fibre Bragg gratings inscribed point-by-point by infrared femtosecond laser. Electronics Letters. 2005;**41**(4):176-178

[38] Jovanovic N, Åslund M, Fuerbach A, Jackson SD, Marshall GD, Withford MJ. Narrow linewidth, 100 W cw Yb 3⁺-doped silica fiber laser with a point-by-point Bragg grating inscribed

directly into the active core. Optics Letters. 2007;**32**(19):2804-2806

[39] Jovanovic N, Fuerbach A, Marshall GD, Withford MJ, Jackson SD. Stable high-power continuous-wave Yb 3⁺-doped silica fiber laser utilizing a point-by-point inscribed fiber Bragg grating. Optics Letters. 2007;**32**(11):1486-1488

[40] Aslund ML, Jovanovic N, Canning J, Jackson SD, Marshall GD, Fuerbach A, et al. Rapid decay of type-II femtosecond laser inscribed gratings within-switched Yb-doped Fiber lasers. IEEE Photonics Technology Letters. 2010;**22**(7):504-506

[41] Mihailov SJ, Smelser CW, Lu P, Walker RB, Grobnic D, Ding H, et al. Fiber Bragg gratings made with a phase mask and 800-nm femtosecond radiation. Optics Letters. 2003;**28**(12): 995-997

[42] Mihailov SJ, Smelser CW, Grobnic D, Walker RB, Lu P, Ding H, et al. Bragg gratings written in all-SiO₂ and Ge-doped core fibers with 800-nm femtosecond radiation and a phase mask. Journal of Lightwave Technology. 2004;**22**(1):94

[43] Grobnic D, Walker RB, Mihailov SJ, Smelser CW, Lu P. Bragg gratings made in highly nonlinear bismuth oxide fibers with ultrafast IR radiation. IEEE Photonics Technology Letters. 2010; 22(2):124-126

[44] Grobnic D, Mihailov SJ, Smelser CW. Femtosecond IR laser inscription of Bragg gratings in single-and multimode fluoride fibers. IEEE Photonics Technology Letters. 2006;**18**(21–24):2686-2688

[45] Grobnic D, Mihailov SJ, Smelser CW, Ding H. Sapphire fiber Bragg grating sensor made using femtosecond laser radiation for ultrahigh temperature applications. IEEE Photonics Technology Letters. 2004;**16**(11): 2505-2507

[46] Mihailov SJ, Grobnic D, Huimin D, Smelser CW, Broeng J. Femtosecond IR laser fabrication of Bragg gratings in photonic crystal fibers and tapers. IEEE Photonics Technology Letters. 2006;**18**(17–20):1837-1839

[47] Voigtländer C, Becker RG, Thomas J, Richter D, Singh A, Tünnermann A, et al. Ultrashort pulse inscription of tailored fiber Bragg gratings with a phase mask and a deformed wavefront [invited]. Optical Materials Express. 2011;**1**(4):633-642

[48] Voigtländer C, Krämer RG, Thomas JU, Richter D, Tännermann A, Nolte S. Variable period change of femtosecond written fiber Bragg gratings with a deformed wavefront. In: MATEC Web of Conferences. Vol. 8. EDP Sciences; 2013. p. 06013

[49] Hnatovsky C, Grobnic D, Mihailov SJ. Through-the-coating femtosecond laser inscription of very short fiber Bragg gratings for acoustic and high temperature sensing applications. Optics Express. 2017;**25**:25435-25446

[50] Hnatovsky C, Grobnic D, Mihailov SJ. Nonlinear photoluminescence imaging applied to femtosecond laser manufacturing of fiber Bragg gratings. Optics Express. 2017;**25**:14247-14259

[51] Bernier M, Trépanier F, Carrier J, Vallée R. Efficient writing of Bragg gratings through the coating of various optical fibers. In: Advanced Photonics. Optical Society of America; 2014, paper BM2D.3

[52] Bernier M, Trépanier F, Carrier J, Vallée R. High mechanical strength fiber Bragg gratings made with infrared femtosecond pulses and a phase mask. Optics Letters. 2014;**39**:3646-3649

[53] Habel J, Boilard T, Messaddeq Y, Trépanier F, Bernier M. Flexible phase-mask writing technique of robust femtosecond FBG for distributed

sensing. In: Advanced Photonics 2018 (BGPP, IPR, NP, NOMA, Sensors, Networks, SPPCom, SOF), OSA Technical Digest (online). Optical Society of America; 2018, paper BM3A.4

[54] Habel J, Boilard T, Frenière J-S, Trépanier F, Bernier M. Femtosecond FBG written through the coating for sensing applications. Sensors. 2017;**17**(11):2519

[55] Li SY, Ngo NQ, Tjin SC, Shum P, Zhang J. Thermally tunable narrow-bandpass filter based on a linearly chirped fiber Bragg grating. Optics Letters. 2004;**29**(1):29-31

[56] Ohn MM, Alavie AT, Maaskant R, Xu MG, Bilodeau F, Hill KO. Dispersion variable fibre Bragg grating using a piezoelectric stack. Electronics Letters. 1996;**32**(21):2000-2001

[57] Perlin VE, Winful HG. Nonlinear pulse switching using cross-phase modulation and fiber Bragg gratings. IEEE Photonics Technology Letters. 2001;**13**(9):960-962

[58] Melloni A, Chinello M, Martinelli M. All-optical switching in phase-shifted fiber Bragg grating. IEEE Photonics Technology Letters. 2000;**12**(1):42-44

[59] Euser T. Ultrafast optical switching of photonic crystals [PhD thesis]. University of Twente; 2007

[60] Notomi M. Manipulating light with strongly modulated photonic crystals. Reports on Progress in Physics. 2010;**73**(9):096501

[61] Boyd RW. Nonlinear Optics. Third Edition. ISBN 978-0-12-369470-6. Oxford, UK: Published by Academic Press/Elsevier, Inc.; 2008

[62] Sanghera JS, Shaw LB, Aggarwal ID. Chalcogenide glass-fiber-based mid-IR sources and applications. IEEE Journal

of Selected Topics in Quantum Electronics. 2009;**15**(1):114-119

[63] Othonos A. Probing ultrafast carrier and phonon dynamics in semiconductors. Journal of Applied Physics. 1998;**83**(4):1789-1830

[64] Reif J, Schmid RP, Schneider T. Femtosecond third-harmonic generation: Self-phase matching through a transient Kerr grating and the way to ultrafast computing. Applied Physics B. 2002;**74**(7–8):745-748

[65] Deeg FW, Stankus JJ, Greenfield SR, Newell VJ, Fayer MD. Anisotropic reorientational relaxation of biphenyl: Transient grating optical Kerr effect measurements. The Journal of Chemical Physics. 1989;**90**(12):6893-6902

[66] Zoweil H, Lit JW. Bragg grating with periodic non-linearity as optical switch. Optics Communications. 2002; **212**(1):57-64

[67] Laniel JM, Bélanger N, Villeneuve A. Nonlinear switching in a Bragg grating with periodic χ (3). In: Conference on Lasers and Electro-Optics. Optical Society of America; 2007. p. JWA53

[68] Furtado Filho AFG, de Sousa JRR, de Morais Neto AF, Menezes JWM, Sombra ASB. Periodic modulation of nonlinearity in a fiber Bragg grating: A numerical investigation. Journal of Electromagnetic Analysis and Applications. 2012;**4**(02):53-59

[69] Ashcroft NW, Mermin N. Solide State Physics. Stamford: Thomson Learning; 1976. p. 13, 26, 80, 119 and 127

[70] Euser TG, Vos WL. Spatial homogeneity of optically switched semiconductor photonic crystals and of bulk semiconductors. Journal of Applied Physics. 2005;**97**:043102

[71] Lipson M. Guiding, modulating, and emitting light on silicon—Challenges and opportunities. Journal of Lightwave Technology. 2005;**23**:4222

[72] van Driel HM. Kinetics of high-density plasmas generated in Si by 1.06- and 0.53μ m picosecond laser pulses. Physical Review B. 1987;**35**:8166

[73] Sabbah AJ, Riffe DM. Femtosecond pump-probe reflectivity study of silicon carrier dynamics. Physical Review B. 2002;**35**:165217

[74] Eichler HJ, Massmann F. Diffraction efficiency and decay times of free-carrier gratings in silicon. Journal of Applied Physics. 1982;**53**(4):3237-3242

[75] Sivan Y, Ctistis G, Yüce E, Mosk AP. Femtosecond-scale switching based on excited free-carriers. Optics Express. 2015;**23**(12):16416-16428

[76] Käding OW, Skurk H, Maznev AA, Matthias E. Transient thermal gratings at surfaces for thermal characterization of bulk materials and thin films. Applied Physics A: Materials Science & Processing. 1995;**61**(3):253-261

[77] Johnson JA, Maznev AA, Bulsara MT, Fitzgerald EA, Harman TC, Calawa V, et al. Phase-controlled, heterodyne laser-induced transient grating measurements of thermal transport properties in opaque material. Journal of Applied Physics. 2012;**111**(2):023503

[78] Graebner JE. Measurement of thermal diffusivity by optical excitation and infrared detection of a transient thermal grating. Review of Scientific Instruments. 1995;**66**(7):3903-3906

[79] Stepanov S. Dynamic population gratings in rare-earth-doped optical fibres. Journal of Physics D: Applied Physics. 2008;**41**(22):224002

[80] Canning J, Sceats MG, Inglis HG, Hill P. Transient and permanent gratings in phosphosilicate optical fibers

produced by the flash condensation technique. Optics Letters. 1995;**20**(21): 2189-2191

[81] Canning J, Sceats MG. Transient gratings in rare-earth-doped phosphosilicate optical fibres through periodic population inversion. Electronics Letters. 1995;**31**(7):576-577

[82] Akin O, Dinleyici MS. An all-optical switching based on resonance breaking with a transient grating. Journal of Lightwave Technology. 2010;**28**(23): 3470-3477

[83] Akin O, Dinleyici MS. Demonstration of pulse controlled all-optical switch/modulator. Optics Letters. 2014;**39**(6):1469-1472

[84] Sivan Y, Rozenberg S, Halstuch A, Ishaaya AA. Nonlinear wave interactions between short pulses of different spatio-temporal extents. Scientific Reports. 2016;**6**

[85] Khurgin JB, Sun G, Chen WT, Tsai WY, Tsai DP. Ultrafast thermal nonlinearity. Scientific Reports. 2015;**5**: 17899

[86] Yu Z, Margulis W, Tarasenko O, Knape H, Fonjallaz PY. Nanosecond switching of fiber Bragg gratings. Optics Express. 2007;**15**(22):14948-14953

[87] Sivan Y, Pendry JB. Broadband time-reversal of optical pulses using a switchable photonic-crystal mirror. Optics Express. 2011;**19**:14502-14507

[88] Sivan Y, Rozenberg S, Halstuch A. Coupled-mode theory for electromagnetic pulse propagation in dispersive media undergoing a spatiotemporal perturbation: Exact derivation, numerical validation, and peculiar wave mixing. Physics Review B. 2016;**93**:144303

[89] Karenowska A, Gregg J, Tiberkevich V, Slavin A, Chumak A, Serga A, et al.

Oscillatory energy exchange between waves coupled by a dynamic artificial crystal. Physical Review Letters. 2012; **108**:015505

[90] Sivan Y, Pendry JB. Time reversal in dynamically tuned zero-gap periodic systems. Physical Review Letters. 2011; **106**:193902-1-193902-4

[91] Sivan Y, Pendry JB. Theory of wave-front reversal of short pulses in dynamically tuned zero-gap periodic systems. Physical Review A. 2011;**84**: 033822-1-033822-13

[92] Chumak A, Tiberkevich V, Karenowska A, Serga A, Gregg J, Slavin A, et al. All-linear time-reversal by a dynamic artificial crystal. Nature Communications. 2010;**1**:141

[93] Wu AQ, Chowdhury IH, Xu X. Femtosecond laser absorption in fused silica: Numerical and experimental investigation. Physical Review B. 2005; **72**(8):085128

[94] Shamir A, Ishaaya AA. Effect of femtosecond photo-treatment on inscription of fiber Bragg gratings. Optics Letters. 2016;**41**(4):765-768

[95] Shamir A, Halstuch A, Sivan Y, Ishaaya AA. Ns-duration transient Bragg gratings in silica fibers. Optics Letters. 2017;**42**(22):4748-4751

[96] Combis P, Cormont P, Gallais L, Hebert D, Robin L, Rullier JL. Evaluation of the fused silica thermal conductivity by comparing infrared thermometry measurements with two-dimensional simulations. Applied Physics Letters. 2012;**101**(21):211908

Vital Sign Measurement Using FBG Sensor for New Wearable Sensor Development

Shouhei Koyama and Hiroaki Ishizawa

Abstract

In this study, we measured the vital signs of a living body using an FBG sensor by installing it at a pulsation point such as the radial artery. We developed a biological model to demonstrate the capability of an FBG sensor. The FBG sensor signal was found to correspond to the changes in diameter of the artery caused by the pressure of the blood flow. Vital signs such as pulse rate, respiratory rate, stress load, and blood pressure were calculated from the FBG sensor signal. While pulse rate and respiration rate were calculated by peak detection of FBG sensor signal. Blood pressure was calculated from the waveform shape of one beat of the FBG sensor signal by PLS regression analysis. All vital signs were calculated with high accuracy. The study helps establish that these vital signs can be calculated continuously and simultaneously. Considering that an FBG sensor can detect a strain with high sensitivity using a small optical fiber, it is expected to be adopted widely as a novel wearable vital sign sensor.

Keywords: FBG sensor, pulse rate, respiration rate, blood pressure, wearable sensor

1. Introduction

In Japan, there is surge in demand for medical care of the elderly as their population continues to increase [1]. This is causing a serious concern especially considering the prevailing shortage of medical staff. Meanwhile, since the Tokyo Olympic Games will be held in 2020, there is a high need for self-healthcare management among healthy people. A simple home health system to monitor the vital signs in elderly people is becoming an absolute necessity, as there is increasing demand for their self-health management on a daily basis. Vital signs are fundamental indicators of human health. These indicators include heart rate, respiration rate, blood pressure, body temperature, level of consciousness.

In order to meet such needs, wearable sensors are being developed by manufacturers to monitor vital signs [2–4]. These sensors are glasses or wristwatch type, they have a characteristic that can measure vital signs continuously. Most of these sensors are of photoelectric pulse wave type measuring the changes in light absorption caused by hemoglobin in blood vessels. These sensors are compact, portable, and easy to install on a human body. However, there are a few issues with these sensors: moisture noise caused by perspiration, skin damage due to the probe pressure [5], and dependence of signal strength on probe mounting position [6]. In addition, people have psychologically stressful for people who do not use wrist watches or eyeglasses from attaching these type wearable sensor. Many photoelectric pulse wave sensors can measure only the pulse rate and cannot measure blood pressure. The currently used measuring many sphygmomanometers are of stationary type and therefore cannot be carried by hand. Accordingly, they are not suitable for home use and continuous monitoring.

The FBG sensor is an optical fiber type highly accurate strain sensor. The FBG sensor has a feature that a plurality of sensors can be installed with one inter-rogator, the optical fiber length is 1 km or more. From these features, FBG sensors are used in building and civil engineering fields. Tam et al. have introduced FBG sensors in railway rail monitoring systems [7, 8]. There are research studies reporting measurement of vital signs using FBG sensors [9–11]. Furthermore, since the sensor part is an optical fiber, it can be introduced into a textile product [12]. Therefore, the FBG sensor is introduced into the wristband or the sleeve of the shirt, and the sensor can be installed on the living body simply by wearing the textile product.

The authors propose that the FBG sensor is installed to the pulsation point of the skin surface and the vital sign can be calculated from the measured signal. The vital signs such as pulse rate, respiratory rate, stress load, and blood pressure are calcu-lated from the measured signal of FBG sensor. In this paper, the details of the strain signal measured at a pulsation point of a human body with the FBG sensor, method of calculation from the measured signal, and measurement accuracy for each vital sign are described.

2. FBG sensor system

An FBG sensor system is composed of an interrogator part with a light source and a detector, and a sensor part with an optical fiber. The schematic and the specifications of the FBG sensor system used in this study are shown in **Figure 1** and **Table 1**, respectively. We used the FPG interrogator system, named PF25-S01 (Nagano Keiki Co., Ltd.) [13]. This interrogator is equipped with an ASE light source that emits near infrared light with a wavelength of 1525–1570 nm that passes through the core of the optical fiber.

In the FBG sensor, a diffraction grating is formed when the refractive index is periodically changed along the axis of the core of the optical fiber. The FBG sensor reflects only a specific wavelength corresponding to the interval period of the dif-fraction grating. The wavelength of the reflected light from the FBG is called Bragg wavelength that follows the Eq. (1),

$$\lambda_{Bragg} = 2n_{eff}\Lambda \tag{1}$$

where, λ_{Bragg} is the Bragg wavelength, n_{eff} is the effective refractive index of the grating portion, and Λ is the grating interval. Since the effective refractive index is constant during a measurement, the Bragg wavelength changes accordingly as the lattice spacing changes. Therefore, when the Bragg wavelength varies, the lattice spacing changes due to the strain in the sensor part. Any distortion applied to the sensor section is detected based on this principle.

The Bragg wavelength reflected by the sensor portion passes through a circulator and is directed to a detection device that is a Mach-Zehnder interferometer. The optical path difference of the interferometer is approximately 5 mm. The homodyne detection method using the Mach-Zehnder interferometer detects the shift length of Bragg wavelength as interference phase shift [14, 15]. The Mach-Zehnder inter-ferometer provides the light outputs through three detectors.

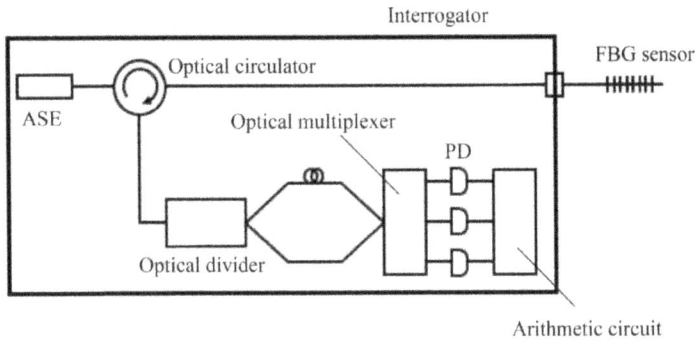

Figure 1.
FBG sensor system.

Interrogator		Optical fiber	
Size D × W × H (mm)	230 × 330 × 100	Material	Silica glass (Core: Ge)
Weight (kg)	4	Mode	Single mode
Light source	ASE	Fiber diameter (μm)	250
Wavelength (nm)	1525–1570	Cladding diameter (μm)	145
Power (mW)	30	Core diameter (μm)	10.5
Sampling rate (kHz)	10	Detection range (nm)	1550 ± 0.5
Detector	InGaAs PIN PD		
Wavelength resolution (pm)	±0.1		

Table 1.
Specification of FBG sensor system.

In each detector, the detected light is photo-electrically converted into an electric signal that is further converted to a digital signal by an analog/digital converter. Subsequently, the phase angle is demodulated, and a wavelength shift (proportional to the displacement and distortion) is calculated. The method has an advantage in that the resolution of wavelength measurement is finer compared with other methods. By this method, the pressure of the FBG sensor part is detected by measuring the displacement in Bragg wavelength.

3. Relationship between human body signal and FBG sensor signal

Experiments were performed using a biological model (**Figure 2**) to study the signals measured by an FBG sensor attached to a living body. A piston was employed to simulate the heart, and a blood-mimicking fluid (manufactured by CIRS) was used to simulate the blood. The movement of the piston was controlled to set the flow rate of the pseudo blood passing through a 500-mm-long acrylic pipe (inner diameter 8 mm) that simulated a blood vessel. For a phantom biological model, Flow Phantom (Supertech, Inc., ATS 524), was used. A flow phantom, made of a rubber material, was used to simulate the resilience of a living body (artificial skin), and it was provided with a hole of 8 mm diameter located at a depth of 15 mm from the top surface. A pipe of 8 mm inner diameter, made of vinyl

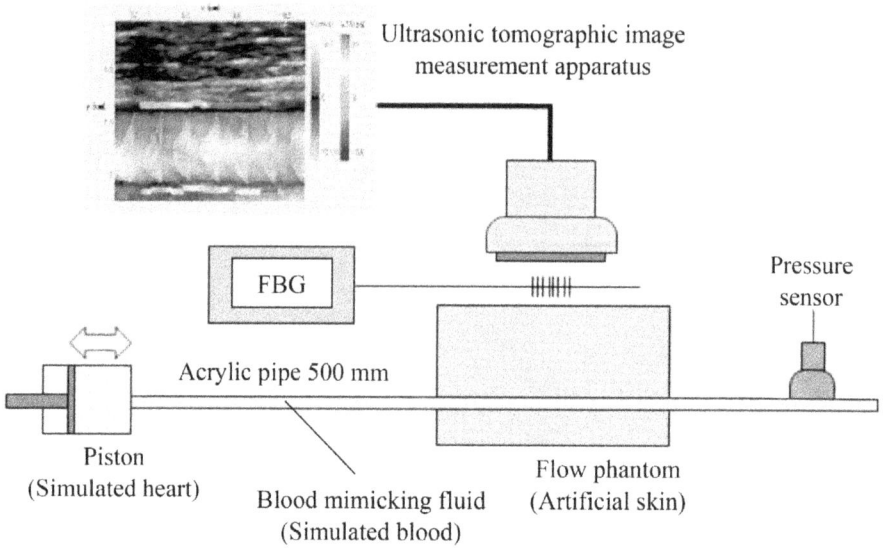

Figure 2.
Schematic of a biological mode.

chloride, and a sensor to measure the pressure of the pseudo blood were installed at the rear of the flow phantom. The pseudo blood, discharged from the piston, passed through the acrylic pipe, flow phantom, and vinyl chloride pipe in this order. An FBG sensor was attached on top of the flow phantom perpendicular to the direction of flow of the pseudo blood. The sensor part of an ultrasonic tomographic image measurement apparatus, installed parallel to the direction of flow of the pseudo blood and covering the FBG sensor as shown in **Figure 2**, captures the image of the inner details of the flow phantom [16].

During the flow of the pseudo blood, the changes in the inner diameter of the pseudo artery were measured by the FBG sensor and the tomographic apparatus. The FBG sensor signal for a flow rate of 30 mL of pseudo blood in 0.5 s, the diameter of the simulated artery from the tomographic image, and the result of the pressure gauge are shown in **Figure 3**. It is evident that the FBG sensor signal is closely similar to the diameter of the simulated artery and the pressure of the fluid. In addition, **Figure 4** shows the FBG sensor signal for various conditions of the pressure of

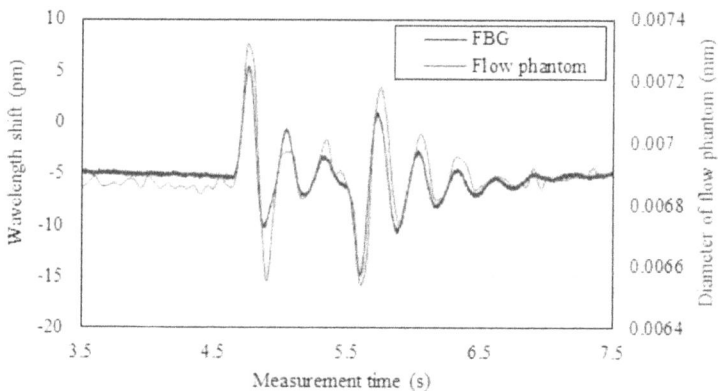

Figure 3.
Result of flow phantom diameter and FBG sensor signal.

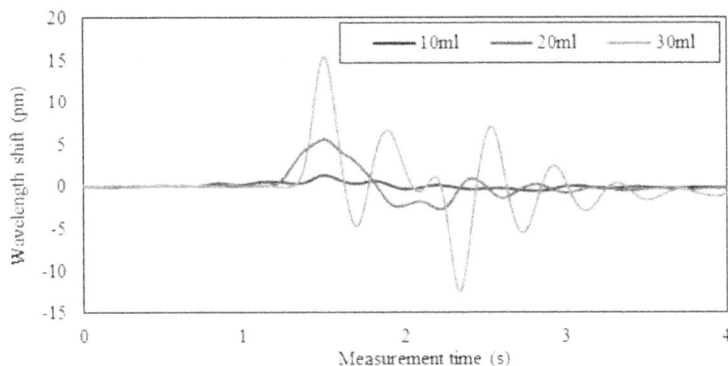

Figure 4.
Result of FBG sensor signal while changing the pressure.

the pseudo blood. It is evident that the larger the pressure of the simulated blood is, the larger the amplitude of the FBG sensor becomes. In this way, it was confirmed that the FBG sensor signal could measure the variations in the artery diameter caused by the blood pressure. In other words, as the strain at the pulsation point is changed by the pressure of the blood flow, the magnitude of the strain change is measured by the FBG sensor; this observation indicates that the FBG sensor signal contains information on blood pressure.

4. Relationship of heartbeat and FBG sensor signal

In Section 3, it was found that the variation in diameter of the blood vessel could be measured by the FBG sensor. In this section, we discuss the relation between the FBG sensor signal and heartbeat by installing the FBG sensor at a pulsation point of the subject.

The FBG sensor was installed perpendicular to the direction of blood flow in the radial artery at the wrist of the subject. An electrocardiograph (Nihon Kohden Corp., PVM-2701) was installed at the chest of the subject, and the electrocardiogram was measured simultaneously. The subject was male in their twenties, their posture at the time of measurement was supine, the sampling rate was 10 kHz, and the duration of measurement was 30 s. The peak-to-peak interval (PPI) of the FBG sensor signal and the R-to-R interval (RRI) of the electrocardiogram were calculated.

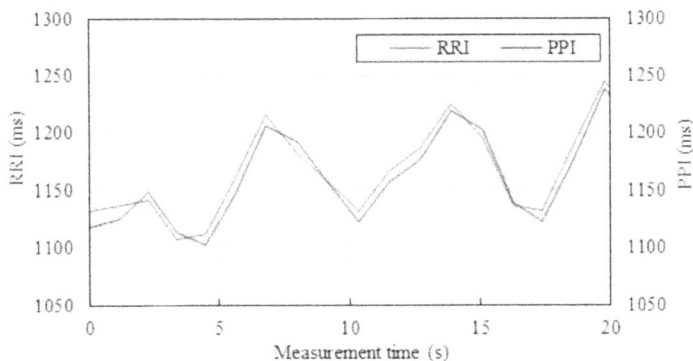

Figure 5.
Results of RRI and PPI.

The result of the RRI and PPI of the subject is shown in **Figure 5**, where the horizontal axis is the measurement time, and the vertical axes are the time intervals of PPI and RRI [17]. The heart rates of subject was ~51 times per min. It is evident that the RRI and PPI plots are almost identical for the subject. In other words, the FBG sensor signal corresponds to the heartbeat vibration as it represents the variation in the diameter of the arterial blood vessel caused by the flow rate (or pressure) that in turn is related to the heartbeat.

5. Vital sign calculation from peak in FBG sensor signal

5.1 Calculation of pulse rate

Since the FBG sensor corresponds to heartbeat vibration, the vital sign can be calculated from the FBG sensor signal. In this section, the pulse rate is calculated from the FBG sensor signal. An FBG sensor was installed perpendicular to the direction of blood flow in the radial artery at the left wrist of the subject. In order to measure the reference pulse rate, an electronic sphygmomanometer was installed at the upper arm of the subject. The measurements using the FBG sensor and the electronic blood pressure were performed at the same time. The peak interval (PPI) of the FBG sensor signal was measured, and the pulse rate per min was calculated from Eq. (2).

$$PR_{cal}(times/minutes) = PPI/60 \qquad (2)$$

where, PR_{cal} is calculated pulse rate, and PPI is the peak interval. Three subjects were measured in supine position, and the measurement conditions were the same as those for the experiment on heartbeat discussed in Section 4.

A scatter diagram between the reference pulse rate and the pulse rate calculated from the FBG sensor signal is shown in **Figure 6**. The correlation coefficients for

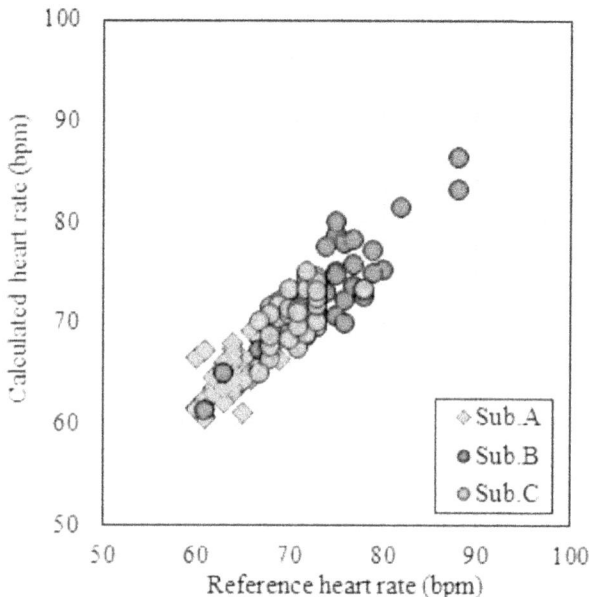

Figure 6.
Result of calculated pulse rate from FBG sensor signal.

the subjects A, B, and C were observed to be 0.67, 0.87, and 0.56, respectively, while the respective measurement accuracies were 2.1, 2.6, and 1.9 bpm. These results indicate that the pulse rate could be measured from the FBG sensor signal with high accuracy. Thus, it is evident that the pulse rate can be calculated if the peak of the FBG sensor signal is detected accurately.

5.2 Calculation of respiratory rate

The pulse rate was calculated from the FBG sensor signal in Section 5.1. This section describes the measurement and calculation of respiration rate. When a person breathes, a physiological phenomenon called respiratory dynamic arrhythmia occurs, whereby the pulse rate rises during inspiration and decreases during expiration. In other words, the PPI decreases during inspiration, and increases during expiration. Therefore, when breathing is repeated, the PPI cycles up and down, and thus, the period of a breath cycle can be deduced from a PPI cycle. The respiratory rate per min is calculated from Eq. (3),

$$RR_{cal} = PPI_{cyc}/60 \tag{3}$$

where, RR_{cal} is calculated respiration rate, and PPI_{cyc} is the cycle of PPI.

Experiments on breathing rate were conducted with three subjects. To measure the reference respiratory rate, a medical face mask attached with a temperature data logger (Ishikura Shoten Co., Ltd., SK-L200 THII) was used. The purpose of the face mask was to prevent the breath from leaking out. Breath temperature is known to be higher than the atmospheric temperature. At the time of inspiration, the temperature of the atmosphere is measured with a data logger. On the contrary, as the breath temperature was measured at exhalation, the value was relatively higher. Thus, a constant periodic temperature change was measured every time the subject breathes. One cycle of this temperature change was used as reference respiration time, and the reference respiratory rate was calculated from Eq. (3). The FBG sensor signal was measured under the same conditions as those for the experiment on pulse rate presented in Section 5.1. The subjects were in sitting posture at the time of measurement.

The results of the temperature data logger and PPI from the FBG sensor signal are shown in **Figure 7**. It is evident that both curves are very similar to each other. In addition, it could be confirmed that the cycle varies depending on the number of breaths. The respiratory rates calculated from the temperature data logger and the FBG sensor signal are plotted against one another in **Figure 8** for the three

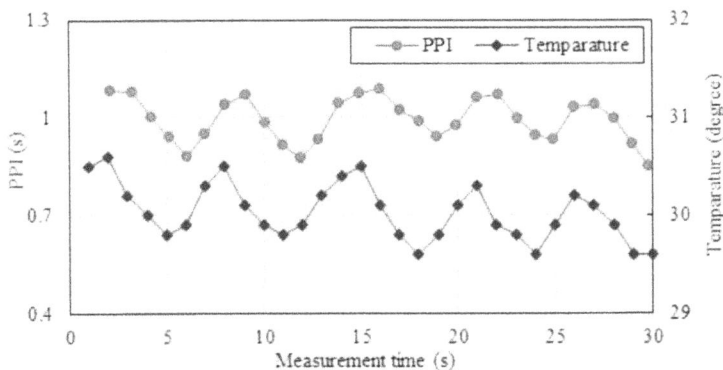

Figure 7.
Results of the PPI from the temperature data logger and the FBG sensor.

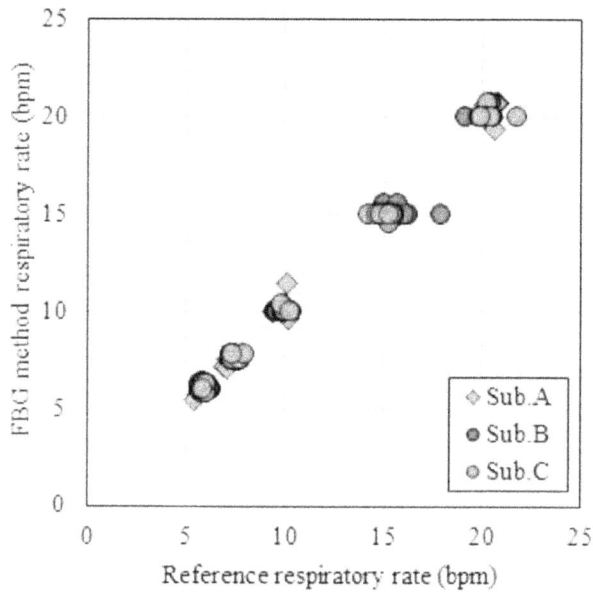

Figure 8.
Result of calculating respiratory rate.

subjects. The measurement accuracies observed for the three subjects were 0.4, 0.6, and 0.4 per min, which were considered to be reasonably good. The high measurement accuracy was observed even for different respiration rates of the same subject and for different subjects. The change in pulse rate (change in PPI interval) due to respiratory dynamic arrhythmia was very small; however, since the sampling rate of the FBG sensor was 10 kHz, it is considered that the calculated respiratory rate was accurate. This measurement method can calculate a respiratory rate in the range of 6–10 bpm; therefore, it is suitable for measurement of slow breathing (~12 bpm or less). It is evident from the above results that the high accuracy of measurement of respiratory rate is attributable to the high sampling rate.

6. Calculating of blood pressure from the waveform of FBG sensor signal

6.1 Waveform of the FBG sensor signal

In this section, the blood pressure is calculated from the waveform of the FBG sensor signal. As shown in Section 3, the FBG sensor signal is measured representing the pressure of the blood flow that causes a change in the diameter of the blood vessel. Pulsation is a distortion that causes an arterial distortion on the skin surface. Therefore, information on blood pressure is considered to be present in the FBG sensor signal from which a distortion is measured.

A signal measured with a general photoelectric pulse wave sensor is a volume pulse wave signal indicating the volume of blood. A signal obtained by second derivative of the volume pulse wave signal is an acceleration plethysmogram. The basic shape of acceleration plethysmogram includes five peaks [18]. The A-wave to the E-wave are called initial systolic positive wave, initial contraction negative wave, mid-systole re-elevation wave, post-contraction descent wave, and expansion initial positive wave, respectively. Therefore, an acceleration pulse wave contains information on systole and diastole of the heart. The first derivative signal of the

FBG sensor is similar to the acceleration pulse wave in shape [13]. Since an FBG sensor signal indicates the displacement of the Bragg wavelength due to strain, time change of the volume pulse wave is measured. Therefore, the first derivative signal of the FBG sensor is similar to the second derivative signal of the volume pulse wave signal. Furthermore, since the first derivative waveform of an FBG sensor signal includes information on the systole and diastole of the heart, the blood pressure can be calculated from this waveform.

6.2 Calculation method of blood pressure from the FBG sensor signal

When blood pressure was calculated from FBG sensor signal, PLS regression analysis, which is a widely known multivariate analysis among others, was used. The PLS regression analysis can construct calibration curves from the explanatory and objective variables. At this time, it is a feature to construct a calibration curve on the premise that an explanatory variable and an objective variable contain errors. The explanatory variable is the FBG sensor signal waveform, and the objective variable is the blood pressure measured simultaneously by the electronic sphygmomanometer. The FBG sensor signal is processed through the following steps [19].

1. The FBG sensor signal is processed with a band pass filter of 0.5–5 Hz.

2. The 'A' peak ('A' wave) is detected from the signal waveform of the band pass filter.

3. The signal is divided for each detected peak.

4. The divided signals are averaged.

5. At the vertical axis of the averaged signal, the first point (peak of 'A' wave) is normalized to "1", and the lowest point (peak of 'B' wave) is set to "0".

6. The horizontal axis of all FBG sensor signal waveforms processed up to the 5th term is cut out in the shortest time, and the length of the horizontal axis is unified.

The step 1 of the signal processing a range for covering a signal with a pulse rate of 30–300 times/min. In step 4 of the signal processing, considering that the measurement time of the electronic sphygmomanometer is ~30 s, the average is calculated for the pulse wave signals measured within that time. The step 5 of signal processing is to cancel the vertical axis fluctuations caused by pressure while installing the FBG sensor in humans. The step 6 of signal processing is to cancel the pulse rate fluctuations caused by respiratory sinus arrhythmia.

The FBG sensor signal processed through the aforementioned signal processing steps is used as an explanatory variable, the blood pressure value of the electronic sphygmomanometer measured simultaneously is used as a target variable, and a calibration curve is constructed by PLS regression analysis. The newly measured FBG sensor signal is substituted into this calibration curve to calculate the blood pressure.

6.3 Experimental result of calculating the blood pressure

This experiment was performed on three subjects. A schematic of the experimental blood pressure measurement is shown in **Figure 9** [19]. The posture of the subject was supine, and the FBG sensor was installed at the pulsation point of the radial artery of the right wrist. The reference blood pressure value (objective

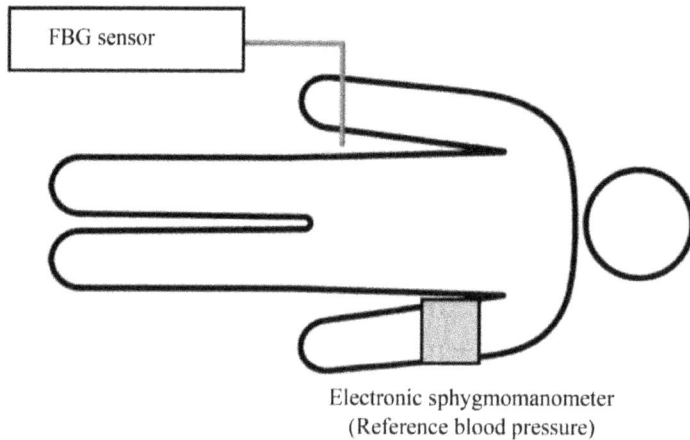

Figure 9.
Experimental image of blood pressure measurement.

variable) was measured simultaneously with the electronic sphygmomanometer installed at the left upper arm. Systolic and diastolic blood pressures were measured with an electronic sphygmomanometer. In the calculation of the systolic blood pressure, the signal-processed FBG sensor signal waveform was used as an explanatory variable, and the systolic blood pressure measured simultaneously with the electronic blood pressure monitor was used as the objective variable. Similarly, in the calculation of the diastolic blood pressure, the same FBG sensor signal waveform and the diastolic blood pressure measured simultaneously with electronic sphygmomanometer were used. The measurement time was 30 s, while the number of measurements was 75 times. Whereas 50 data points were used for construction of calibration curve, the remaining 25 data points were assigned to the calibration curve and used as verification data for blood pressure calculation. The target measurement accuracy was ±5 mmHg.

Table 2 shows the calibration curve construction data sets of systolic and diastolic blood pressures in each subject [19]. A calibration curve for calculating systolic blood pressure or diastolic blood pressure was constructed using the data sets of each subject. **Table 3** shows verification data sets for calculation of systolic and diastolic blood pressures for each subject. The verification data set is substituted into the constructed calibration curve, and systolic and diastolic blood

Subject	Number	Max (mmHg)	Min (mmHg)	Ave (mmHg)
(a) Systolic blood pressure data set				
A	50	125	100	111.3
B	50	136	113	123.1
C	50	111	93	100.9
(b) Diastolic blood pressure data set				
A	50	79	46	62.7
B	50	80	56	68.2
C	50	60	46	53.7

Table 2.
Calibration curve construction data sets.

Subject	Number	Max (mmHg)	Min (mmHg)	Ave (mmHg)
(a) Systolic blood pressure data set				
A	25	131	100	110.1
B	25	138	112	122.2
C	25	110	93	100.4
Average systolic blood pressure in data set (mmHg)				110.9
(b) Diastolic blood pressure data set				
A	25	77	46	60.3
B	25	76	58	68.0
C	25	65	39	53.1
Average diastolic blood pressure in data set (mmHg)				60.5

Table 3.
Verification data set in each subjects.

pressures were calculated. For example, a calibration curve was constructed with 50 data points of the systolic blood pressure of the subject A; the 25 data points of the systolic blood pressure of verification data of the subject A were substituted into the calibration curve, and the systolic blood pressure of subject A was calculated. Similarly, a calibration curve was constructed with 50 data points of the diastolic blood pressure of the subject B in **Table 2**; the 25 data points of validation data of the diastolic blood pressure of the subject B in **Table 3** were substituted into the calibration curve, and the diastolic blood pressure of subject B was calculated.

Figure 10 shows a scatter plot of reference blood pressure and calculated blood pressure during systole and diastole of each subject. **Table 4** shows the results of blood pressure calculation, whereby it was observed that the calculation accuracy of systolic and diastolic blood pressures were ±5 mmHg, and it was calculated with the same blood pressure value as that of a commercially available blood pressure monitor. In the case of the systolic blood pressure, the average value of the verification data for blood pressure calculation was 110.9 mmHg, while the average value

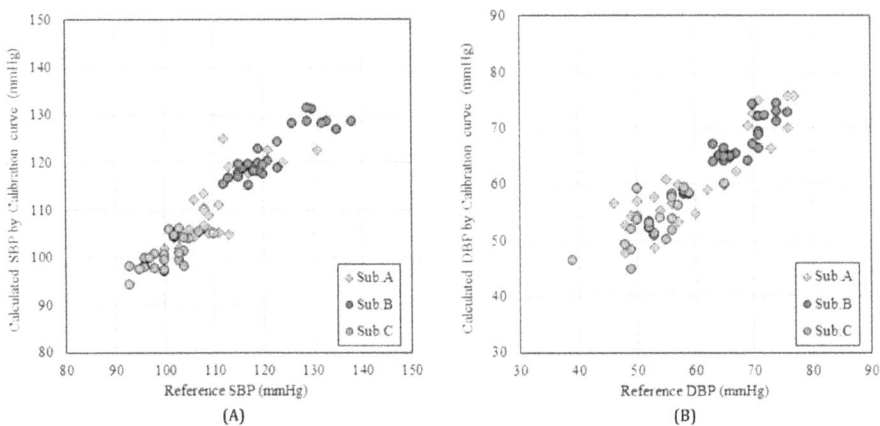

Figure 10.
Scatter plot of reference blood pressure and calculated blood pressure during systolic and diastolic. (A) Calculated result in systolic blood pressure. (B) Calculated result in diastolic blood pressure.

	Subject	A	B	C	Average
Calculating accuracy (mmHg)	Systolic	3.4	2.8	2.6	2.9
	Diastolic	4.1	1.8	2.6	2.8

Table 4.
Results of calculating blood pressure.

of the calculation accuracy was 2.9 mmHg (~2.6%). Similarly, in the diastolic blood pressure, the average value of the verification data for blood pressure calculation was 60.5 mmHg, while the average value of the calculation accuracy was 2.8 mmHg (~4.7%). These results indicate that the calculation accuracy of diastolic blood pressure is lower than that of systolic blood pressure. This is due to step 6 of signal processing, whereby the canceling of the pulse rate fluctuation was performed in order to calculate the FBG sensor signal by the PLS regression analysis. The so-called latter half portion of a single beat of the FBG sensor signal is truncated. There is a peak of expansion initial positive wave in this part, which represents the diastole of the heart. Therefore, it is considered that the deletion of the diastolic information from the FBG sensor signal caused a decline in the calculation accuracy. In other words, in this calculation method, since the negative characteristics of the signal processing is reflected in the result, the blood pressure is calculated from the movement of the heart included in the FBG sensor signal. However, considering that the results of all blood pressure calculations were ±5 mmHg, it is considered that the blood pressure was calculated with high calculation accuracy. Therefore, it is established that the blood pressure can be calculated with high calculation accuracy by constructing the calibration curve by PLS regression analysis of the waveform of the FBG sensor signal. It is evident from this result that it is possible to calculate blood pressure from the same FBG sensor signal in addition to pulse rate, respiratory rate, and stress load.

7. Conclusion

In this paper, the method of calculating each vital sign from the FBG sensor signal was described. An experiment using a biological model demonstrated that the FBG sensor signal was influenced by the change in diameter of the tube through which a fluid (pseudo blood) was allowed to flow by a simulated pressure. This model was replaced with a living body, and a change in the diameter of the artery caused by a change in the flow rate of blood related to the movement of the heart was measured. The FBG sensor signal was measured at a high temporal resolution of 10 kHz; therefore, the pulse rate could be calculated with a high measurement accuracy. Based on the phenomenon of respiratory sinus arrhythmia, the respiration rate could be calculated from the cycle of pulse rate change during expiration and inspiration. On the other hand, the first derivative waveform of the FBG sensor signal was found to be similar to the acceleration pulse wave, which was the second derivative of the volume pulse wave signal; therefore, it was considered that the systolic and diastolic information of the heart was present in the signal. Subsequent to the signal processing of this primary derivative waveform of FBG sensor signal, a calibration curve was constructed by PLS regression analysis for calculating the blood pressure, and accordingly, the blood pressures (systolic and diastolic) were calculated with high accuracy. All these vital signs were calculated from a measured FBG sensor signal. Therefore, considering that only the analysis method is different, it is possible to calculate a plurality of vital signs simultaneously from one measurement signal. Since the FBG sensor signal is continuously measured, the vital signs can also be calculated on a continuous basis. In addition, the signal is measured simply by installing the FBG sensor at the pulsation point. Apart from the wrist, there are other pulsation points such as at the elbow, neck, temple, and ankle of the living body. Therefore, it is possible to select a part of interest that can be easily measured. Since the FBG sensor is an optical fiber, it can be introduced into a textile product. Cover the optical fiber with multiple silk thread in technic of braid. Unevenness is formed on the surface of the optical fiber by this silk thread, and frictional force is occurred. The optical fiber covered with this silk thread is embedded into the knitted fabric by inlay knitting. With this method, the optical fiber can be fixed without slipping in the knitted fabric. If FBG sensor can be introduced into textile products, it can be applied to wristbands, long-sleeved shirts, socks, and various other products. However, there are issues such as miniaturization of the FBG interrogator.

There are two approaches to applying the FBG sensor system in the clinical setting. First method, place the current FBG interrogator in the corner of the room, extend the optical fiber and install the FBG sensor in the hospitalized patient. With this method power supply can be supplied stably. The second method is to develop a compact FBG interrogator that can be attached to the human body. We are developing the 1/12 downsizing prototype FBG interrogators ($74 \times 97 \times 57$ mm, 175 g) using commercially available optical components. This interrogator is battery-powered and wirelessly communicates measurement signals. If electronic circuits etc. are changed from commercial type to specialization type, further size down is predicted. This prototype FBG interrogator can be operated at minimum 5 hours with battery. However, it is necessary to take measures against the temperature caused by heat. In our study, we employed the FBG sensor commonly used in applications ranging from infrastructure to living organisms. While we used only one FBG sensor for all our experiments, the length of the optical fiber was also very short. The currently available commercial FBG interrogators have been specifically developed for use in large-scale applications employing many sensors. Our research is a highly original theme that is entirely different from the current application areas of FBG sensors. An FBG sensor can measure a distortion on a continuous basis with high accuracy and sensitivity, and the sensor section is compact owing to the use of optical fiber. Furthermore, it facilitates measurement of a pulse wave signal by a very simple method of fixing it at a pulsation point. Except for a few issues, the FBG sensors having such characteristics are expected to be used as wearable sensors successfully in future.

Currently, there are very few products that can continuously measure blood pressure without using a cuff. In addition, pulse rate and stress load can be detected with commercially available wearable sensors. However, at the same time, there is no wearable sensor using which both breathing rate and blood pressure can be calculated. This is because of the issues related to the time resolution and detection sensitivity of the measurement signal. In contrast, the FBG sensor can calculate these vital signs with high accuracy. It can be used as a precious blood pressure monitor. Moreover, considering that hospitalized patients need to be monitored on a continuous basis in respect of the vital signs, it would be a major breakthrough if the sensors can be installed in textile products for ease of handling and installation. The above discussions lead to the conclusion that it is possible to use an FBG sensor for monitoring of vital signs with high accuracy.

Acknowledgements

This work was supported by JSPS KAKNHI Grant Number JP16H01805 and the Wearable vital signs measurement system development project at Shinshu University. This research is (partially) supported by the Creation of a development platform for implantable/wearable medical devices by a novel physiological data integration system of the Program on Open Innovation Platform with Enterprises, Research Institute and Academia (OPERA) from the Japan Science and Technology Agency (JST).

Conflict of interest

The authors declare no conflict of interest.

Author details

Shouhei Koyama[1*] and Hiroaki Ishizawa[2]

1 Faculty of Textile Science and Technology, Shinshu University, Japan

2 Institute for Fiber Engineering, Shinshu University, Japan

*Address all correspondence to: shouhei@shinshu-u.ac.jp

IntechOpen

© 2019 The Author(s). Licensee IntechOpen. This chapter is distributed under the terms of the Creative Commons Attribution License (http://creativecommons.org/licenses/by/3.0), which permits unrestricted use, distribution, and reproduction in any medium, provided the original work is properly cited. (cc) BY

References

[1] Cabinet Office, Government of Japan. White Paper on Aging Society [Internet]. 2018. Available from: http://www8.cao.go.jp/kourei/whitepaper/w-2018/zenbun/pdf/1s1s_01.pdf [Accessed: November 15, 2018]

[2] Microsoft Co., Ltd. Glabella: Continuously Sensing Blood Pressure Behavior using an Unobtrusive Wearable Device [Internet]. 2017. Available from: https://www.microsoft.com/en-us/research/wp-content/uploads/2017/09/2017-imwut17-holz_wang-Glabella-Continuously_Sensing_Blood_Pressure_Behavior_using_an_Unobtrusive_Wearable.pdf [Accessed: November 15, 2018]

[3] Apple Inc. Apple Watch [Internet]. 2018. Available from: https://support.apple.com/en-am/HT204666 [Accessed: November 15, 2018]

[4] Seiko Epson Corp. Pulsense [Internet]. 2018. Available from: https://global.epson.com/innovation/core_technology/wearable/vital_sensing.html [Accessed: November 15, 2018]

[5] Wille J, Braams R, van Haren WH, van der Werken C. Pulse oximeter-induced digital injury: Frequency rate and possible causative factors. Critical Care Medicine. 2000;**28**(10):3555-3557

[6] Kellher JF, Ruff RH. The penumbra effect: Vasomotion-dependent pulse oximeter artifact due to probe malposition. Anesthesiology. 1989;**71**(5):787-791

[7] Tam HY, Liu SY, Guan BO, Chung WH, Chan TH, Cheng LK. Fiber Bragg grating sensors for structural and railway applications. In: Proceedings of International Society for Optical Engineering (SPIE). Photonics Asia; 8-12 November 2004; Beijing, China. 2004. pp. 85-97

[8] Wei CL, Lai CC, Liu SY, Chung WH, Ho TK, Tam HY, et al. A fiber Bragg grating sensor system for train axle counting. IEEE Sensors Journal. 2010;**10**(12):1905-1912. DOI: 10.1109/JSEN.2010.2049199

[9] Elsarnagawy TD, Haueisen J, Farrag MA, Ansari SG, Fouad H. Embedded fiber Bragg grating based strain sensor as smart costume for vital signal sensing. Sensor Letters. 2014;**12**(11):1669-1674. DOI: 10.1166/sl.2014.3382

[10] Spillman WB Jr, Mayer M, Bennett J, Gong J, Meissner KE, Davis B, et al. A 'smart' bed for non-intrusive monitoring of patient physiological factors. Measurement Science and Technology. 2004;**15**(8):1614-1620. DOI: 10.1088/0957-0233/15/8/032

[11] Hao J, Jayachandran M, Kng PL, Foo SF, AungAung PW, Cai Z. FBG-based smart bed system for healthcare applications. Frontiers of Optoelectronics. 2010;**3**(1):78-83. DOI: 10.1007/s12200-009-0066-0

[12] Sakaguchi A, Kato M, Ishizawa H, Kimura H, Koyama S. Fabrication of optical fiber embedded knitted fabrics for smart textiles. Journal of Textile Engineering. 2016;**62**(6):129-134. DOI: 10.4188/jte.62.129

[13] Koyama S, Ishizawa H, Sakaguchi A, Hosoya S, Kawamura T. Influence on calculated blood pressure of measurement posture for the development of wearable vital sign sensors. Journal of Sensors. 2017;**2017**:8916596. DOI: 10.1155/2017/8916596

[14] Yoshino T, Sano Y, Ota D, Fujita K, Ikui T. Fiber-Bragg-grating based single axial mode Fabry-Perot interferometer and its strain and acceleration sensing applications. Journal of Lightwave

Technology. 2016;**34**(9):2240-2250.
DOI: 10.1109/JLT.2016.2521440

[15] Todd MD, Johnson GA, Chang CC.
Passive, light intensity-independent
interferometric method for fiber Bragg
grating interrogation. Electronics
Letters. 1999;**35**(22):1970-1971. DOI:
10.1049/el:19991328

[16] Kadowaki H, Hayase T, Funamoto
K, Sone S, Shimazaki T, Jibiki T, et al.
Blood flow analysis in carotid artery
bifurcation by two-dimensional
ultrasonic-measurement-integrated
simulation. Journal of Biomechanical
Science and Engineering. 2014;**10**(1):
14-00266. DOI: 10.1299/jbse.14-00266

[17] Koyama S, Ishizawa H, Hosoya S,
Kawamura T, Chino S. Stress loading
detection method using the FBG
sensor for smart textile. Journal
of Fiber Science and Technology.
2017;**73**(11):276-283. DOI: 10.2115/
fiberst.2017-0042

[18] Takazawa K, Fujita M, Kiyoshi Y,
Sakai T, Kobayashi T, Maeda K, et al.
Clinical usefulness of the second
derivative of a plethysmogram
(acceleration plethysmogram).
Cardiology. 1993;**23**:207-217

[19] Koyama S, Ishizawa H, Fujimoto K,
Chino S, Kobayashi Y. Influence of
individual differences on the calculation
method for FBG-type blood pressure
sensors. Sensors. 2017;**17**:1-48. DOI:
10.3390/s17010048

Chapter 5

The State-of-the-Art of Brillouin Distributed Fiber Sensing

Cheng Feng, Jaffar Emad Kadum and Thomas Schneider

Abstract

The distributed Brillouin sensing technique has been developed rapidly since its first demonstration three decades ago. Numerous investigations on the performance enhancement of Brillouin sensors in respect to spatial resolution, sensing range, and measurement time have paved the way to its industrial and commercial applications. This chapter provides an overview of different Brillouin sensing techniques and mainly focuses on the most widely used one, the Brillouin optical time domain analysis (BOTDA). The history and the development of Brillouin sensing regarding the performance enhancement in various methods and their records will be reviewed, commented, and compared with each other. As well, related sensing errors and limitations will be discussed, together with the corresponding strategies to avoid them.

Keywords: stimulated Brillouin scattering, distributed fiber sensing, nonlocal effects, modulation instability, pump depletion, polarization fading, spatial resolution, dynamic sensing

1. Introduction

Since optical fibers can be used to measure a variety of physical parameters directly or indirectly, sensing in optical fibers has been intensively investigated in the past few decades [1]. Fiber optic sensors can be split into two big categories, that is, point and distributed sensors. The former type, such as the fiber Bragg grating (FBG), measures the physical parameters only at a particular location but with relative high resolution and sensitivity. The detectable range of the latter one, such as optical time-domain reflectometry (OTDR) [2], is relatively large (usually the fiber length itself) and continuous but with only moderate resolution and limited sensitivity [3]. However, numerous investigations have been carried out to design a novel sensor with the advantages of both types, but avoiding their disadvantages.

With a moderate spatial resolution in the range of 1 m, distributed Brillouin sensing offers a more cost-effective solution than applying numerous point sensors in long-range sensing and a more accurate result than traditional distributed sensing methods such as OTDR [4]. Due to the recent success in breaking the limits of numerous sensing parameters such as spatial resolution, sensing range, and measurement time, the Brillouin sensing technique has proved its eligibility in a variety of fields. Thus, it has aroused significant interest in many different applications like civil engineering, oil and gas pipelines leakage monitoring, and other infrastructure surveillance tasks.

2. Basics of SBS

2.1 Spontaneous and stimulated Brillouin scattering

Brillouin scattering, named after the French physicist Léon Brillouin, who theo-retically predicted light scattering by a thermally excited acoustic wave (phonon) in 1922 [5], is one of the most prominent nonlinear effects in optical fibers [6]. For spontaneous Brillouin scattering (SpBS), an incident photon (pump wave) is transformed into a frequency-downshifted scattered photon (Stokes wave) and a phonon (acoustic wave). The angular distribution of the Stokes wave is governed by the laws of momentum and energy conservation, that is,

$$\vec{k_A} = \vec{k_p} - \vec{k_s} \qquad \omega_B = \omega_p - \omega_s \tag{1}$$

where $\vec{k_A}$, $\vec{k_p}$, and $\vec{k_s}$ are the wave vector of acoustic, pump, and Stokes wave, and ω_B, ω_p, and ω_s are their corresponding angular frequencies. Considering the fiber geometry and provided that the phonon frequency is much smaller than that of both photons, an efficient Brillouin scattering only occurs in the backward direction. The Brillouin frequency shift (BFS) $\nu_B = \omega_B/2\pi$ is estimated to be ~11 GHz for typical single mode fibers (SMF) and a pump wavelength in the C-Band of optical telecommunications (around 1550 nm) [7]. As has been shown very recently, forward SBS [8] can be used for sensing applications as well [9, 10]. However, the interaction is governed by transverse acoustic modes, and compared to backward SBS, the effect is rather weak. Since a description of this new field of SBS in sensing would go far beyond the scope of this book chapter, hereafter only backward SBS and its applications are discussed.

The basic origin of SBS is electrostriction, which tends to compress the material in the presence of an electrical field [11]. The superposition of the pump and the counterpropagating Stokes wave modulates the density and hence the refractive index of the optical fiber through electrostriction. Thus, a moving grating (an acous-tic wave) is formed. If the velocity of this moving grating coincides with the speed of sound in the material, the effect is very efficient and it additionally reflects optical power from the pump wave. Due to the Doppler effect, the reflected pump wave is downshifted in frequency by the frequency difference between pump and probe and thus adds power to the Stokes wave. Therefore, a positive feedback loop is established. This transformation from SpBS to SBS can be quantitatively described by the differential equation system (DES) in the propagation direction z as:

$$\frac{dI_p}{dz} = -g_B(\omega)I_pI_s - \alpha I_p$$
$$-\frac{dI_s}{dz} = g_B(\omega)I_pI_s - \alpha I_s \tag{2}$$

where I_p and I_s are the intensity of pump and Stokes wave, α is the fiber attenua-tion, and $g_B(\omega)$ is the SBS gain.

The threshold of SBS is defined by the critical power that characterizes the transformation from SpBS to SBS. However, its definition is rather controversial. Smith et al. first defined it as the input pump power at which the backscattered power equals to the transmitted power at the output [12] and Eq. (3) with the critical gain factor $C_{th} \approx 21$ gives an estimation of this critical value:

$$P_{th} \equiv C_{th} \frac{A_{eff}}{g_B(\omega_B) L_{eff}} \tag{3}$$

where $g_B(\omega_B)$ is the peak SBS gain, $L_{eff} = [1 - \exp(-\alpha L)]/\alpha$ is the effective length with L as the real fiber length, A_{eff} is the effective cross section of the fiber, and P_{th} is the estimated SBS threshold. Later on, this estimation has been revised with different theories and approximations, achieving also different C_{th} values [13–15]. However, instead of a constant, C_{th} has been recently found to be dependent on the fiber length and its parameters [16]. As a typical experimental example, **Figure 1(a)** illustrates the transmitted and the reflected power as a function of the incident power in a 20 km SMF. The dashed black line symbolizes the SBS threshold, beyond which the backscattered power rapidly increases and the output pump power stays almost constant.

2.2 The Brillouin gain

The transferred energy from the pump to the Stokes wave can be regarded as an amplification when it is frequency downshifted to the pump wave by BFS. For typical SMF in the C-Band of optical telecommunications, the lifetime T_B of the phonon involved in the SBS interaction is usually in the magnitude of ~10 ns, which leads to a finite spectral distribution of the SBS gain. The complex SBS gain coefficient, which depicts the evolution of the probe wave as a function of frequency detuning ω between pump and probe wave, is approximated by a Lorentzian shape [17]:

$$g(\omega) = \frac{g_0 P_p}{1 - 2j(\omega - \omega_B)/\Gamma_B} \tag{4}$$

where $\Gamma_B = T_B^{-1}$, P_p is the pump power, g_0 is related to the inherent material Brillouin gain g_p with $g_0 = g_p/A_{eff}$ and A_{eff} is the effective cross section of the fiber, g_p is in the range of 3×10^{-11} m/W to approximately 5×10^{-11} m/W at 1550 nm. The real part of Eq. (4) represents the power amplification of the probe wave, also known as the Brillouin gain spectrum (BGS), and can be expressed as:

$$g_B(\omega) = \frac{g_0 P_p (\Gamma_B/2)^2}{(\omega - \omega_B)^2 + (\Gamma_B/2)^2} \tag{5}$$

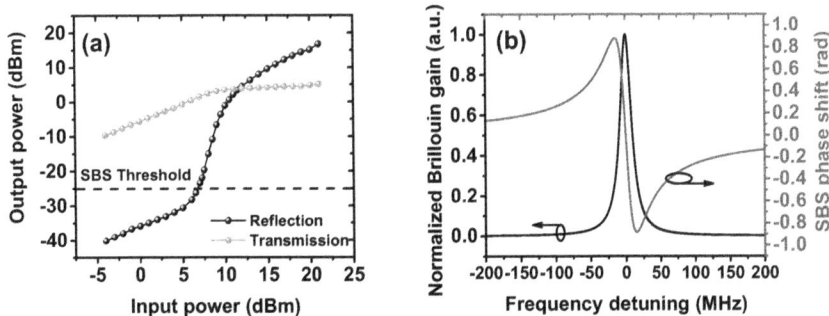

Figure 1.
(a) The transition from SpBS to SBS in a 20 km SMF, from a distinct threshold (here $C_{th} \approx 16$), the reflected stokes power drastically increases, whereas the transmitted pump power stays almost constant; (b) simulated normalized Brillouin gain spectrum (BGS) and Brillouin phase spectrum (BPS) of a 20 km SMF.

The full width at half maximum (FWHM) of the BGS $\Delta \nu_B = \Gamma_B/(2\pi)$, also known as the natural Brillouin linewidth, is estimated to be several tens of MHz for SMF, as shown in **Figure 1(b)** for a 20 km SMF. The Brillouin linewidth can be engineered to several GHz with a broadened pump for applications such as SBS-based filters [18–20] and slow light [21–23] while it can also be narrowed to a record of 3.4 MHz with specific techniques [24–28].

It is also worth to mention that the complex SBS gain changes not only the probe wave in amplitude, but also in phase by its imaginary part. As shown in **Figure 1(b)**, the SBS phase shift has a linear dependence on the frequency detuning in the vicinity of the BFS and zero-phase shift directly at the BFS, that is, the peak Brillouin gain. These properties are important for applications such as SBS-based microwave photonic filters [29] and distributed Brillouin dynamic sensing [30].

2.3 The polarization effect of SBS

Since the state of polarization (SOP) of the probe and pump waves hover randomly in the fiber due to the weak birefringence of SMF and only an averaged interaction can be detected at the output, the dependence of the SBS gain on the SOP has been neglected for long. However, with the rise of the polarization maintaining technique [31], the polarization effect of SBS has been investigated, and it has been found that the SBS efficiency of the pump-probe interaction with arbitrary polarization states is governed by [7]:

$$\eta_{SBS} = \frac{1}{2}(1 + \hat{s} \cdot \hat{p}) = \frac{1}{2}(1 + s_1 p_1 + s_2 p_2 - s_3 p_3) \tag{6}$$

where $\hat{s} = (s_1, s_2, s_3)$ and $\hat{p} = (p_1, p_2, p_3)$ represent the unit vectors of Stokes and pump waves in the Poincaré sphere. As shown in Eq. (6), the SBS efficiencies of a parallel and orthogonal SOP are $1 - s_3^2$ and s_3^2. The SBS efficiency reaches zero when both of the pump and Stokes are orthogonally linear polarized ($s_3 = 0$). This behavior can be used for many different applications from filters [19, 32, 33] via high-resolution spectrum analyzers [34–36] to the generation of THz waves [37, 38].

3. Principle of distributed Brillouin sensing

According to the DES of SBS, the Brillouin gain in each fiber section is accumulated in an SBS interaction between two continuous waves (CW). This accumulation along the fiber leads to a relative high energy conversion at the detector but makes it difficult to distinguish the information of local interactions. Therefore, distributed Brillouin sensing uses other techniques.

3.1 Temperature and strain-dependent Brillouin frequency shift

According to the theory of material science, the velocity of the longitudinal acoustic mode in the fiber depends on material properties such as Young's moduli and the density [39]. This high sensitivity to the temperature and tensile strain makes the BFS also temperature [40] and strain [41] dependent. The linear dependence has been proved and measured in several papers [40–42], as illustrated in **Figure 2** and can be expressed as:

$$\nu_B(T, \varepsilon) - \nu_B(T_0, \varepsilon_0) = C_\varepsilon \cdot \delta\varepsilon + C_T \cdot \delta T \tag{7}$$

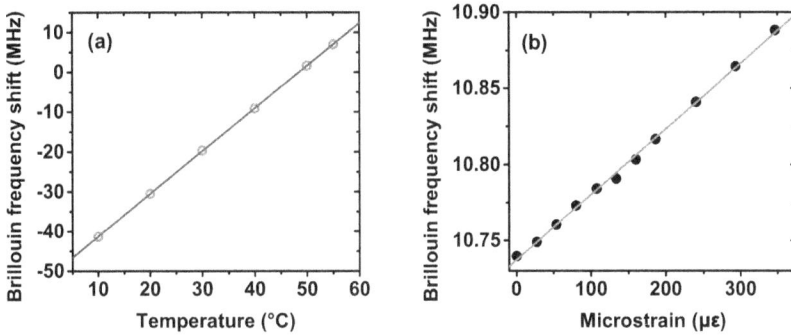

Figure 2.
BFS dependence on (a) temperature and (b) strain in a SMF for a pump wavelength of 1550 nm [42].

where $\nu_B(T, \varepsilon)$ represents the BFS at a temperature T and strain ε, C_T and C_ε are the temperature and strain coefficients. Although both temperature and strain contribute to the BFS shift, the physical difficulty in discriminating the response from these two factors can be solved with specific strategies [43]. For standard SMF, C_T and C_ε are measured to be 1.081 MHz/°C and 42.93 kHz/µε, respectively [42]. The slope of the linearity has also been studied intensively and optimized with different doping concentrations [44].

3.2 Overview of SBS sensing techniques

Since the first demonstration of the most widely used distributed Brillouin sensing scheme in time domain [4], which is now called Brillouin optical time domain analyzer (BOTDA), several different schemes have been proposed and developed with their own advantages and disadvantages.

3.2.1 BOTDA

The principle of BOTDA is based on the Brillouin interaction between a pulsed pump (or probe) wave and a counterpropagating CW probe (or pump) wave. The acoustic wave is generated locally at the point where the pump pulse and the probe CW meet. The energy transfer via the acoustic wave at each position of the fiber under test (FUT) is determined by the frequency detuning between the two signals in comparison to the phonon frequency, that is, the probe wave is amplified when $\nu_p - \nu_s = \nu_a$ and depleted when $\nu_s - \nu_p = \nu_a$ where ν_p, ν_s, and ν_a are the pump, probe, and phonon frequency, respectively. As shown in **Figure 3(a)** as a typical example, local Brillouin gain (or loss) can be translated from the time-dependent to the distance-dependent information according to the round trip relation $z = ct/(2n)$ where z is the fiber position where pump and probe wave interact, t is the propagation time of the pulse, c is the velocity of light in vacuum, and n is the refractive index.

In order to derive the local BFS at each position of the fiber, the reconstruction of the BGS, as illustrated in **Figure 3(b)**, should be carried out by a frequency sweep with every frequency detuning around the phonon frequency. The accurate BFS at each fiber section can be achieved by fitting every measured BGS with the theoretical profile (either Voigt [45] or Lorentzian, dependent on pulse width) as illustrated in **Figure 4(a)**.

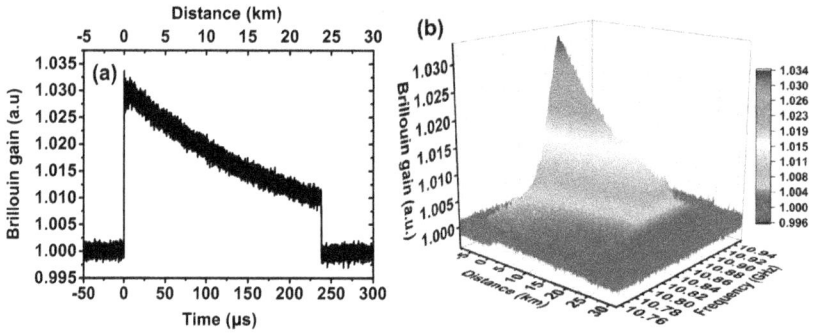

Figure 3.
(a) Evolution of the Brillouin gain for a given frequency detuning; (b) reconstructed 3D BGS along the fiber.

3.2.2 BOTDR

The principle of Brillouin optical time domain reflectometry (BOTDR) is a Brillouin scattering-based OTDR [2]. Different from BOTDA, only a pulsed pump is launched into the FUT from one side of the fiber. Since the time-resolved probe signal is back reflected due to SpBS instead of SBS, the signal power is much weaker than for BOTDA [4] and it is of great importance to apply a coherent detection with a strong local oscillator simultaneously [46]. Since an access to the other fiber end is not necessary, BOTDR is advantageous for some applications. However, besides the weak received signal, it suffers from further disadvantages such as limited spatial resolution of around 1 m, the distortion from Rayleigh backscattering, Fresnel reflection from the connector, and a limited sensing range due to fiber attenuation.

3.2.3 BOCDA

The Brillouin optical correlation domain analyzer (BOCDA) is one of the most recently demonstrated Brillouin sensing techniques. Compared to BOTDA and BOTDR, much higher spatial resolutions down to several millimeters can be achieved [47]. Its principle is based on the interaction of two identically frequency-modulated (FM) counterpropagating CW waves. Similar to the principle of a standing wave (see **Figure 5**), the frequency difference between the counterpropagating pump and probe wave remains constant at specific positions of the fiber, that is, correlation peaks called nodes. Brillouin interactions will take place at these

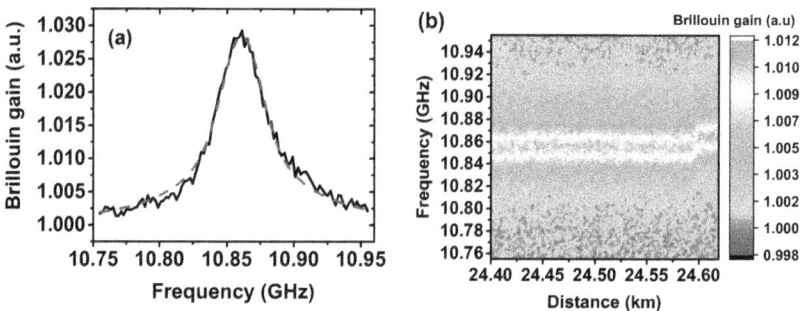

Figure 4.
(a) Measured BGS (solid) at a given fiber section with its Lorentzian fitting (dashed); (b) the reconstructed Brillouin gain mapping with a 20 m long hot spot at the fiber end.

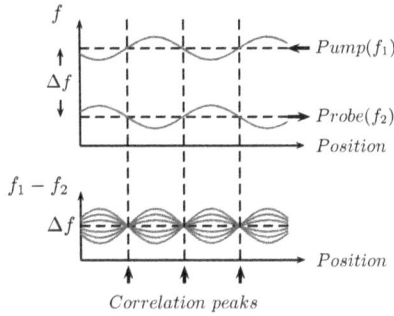

Figure 5.
Schematic explanation of BOCDA [48]. The frequency difference between pump and probe wave is constant at the correlation peaks.

nodes with the frequency difference properly set in the vicinity of the BFS of the FUT [48]. Since the frequency difference varies much faster than the time to excite an acoustic wave, only negligible Brillouin interactions will take place at the other positions. Therefore, unlike a pulse-based scheme such as BOTDA, the spatial resolution of a BOCDA system is usually high and determined by the modulation parameters (amplitude and frequency) written as:

$$\Delta z = \frac{v_g \cdot \nu_B}{2\pi f_m \cdot \Delta f} \tag{8}$$

where v_g is the group velocity, ν_B is the Brillouin gain linewidth, f_m and Δf are the modulation frequency and amplitude of the FM. By sweeping the frequency difference between the pump and probe wave, the BGS of the measured fiber position is scanned.

It is difficult to determine the contribution of the Brillouin interaction from multiple points of the fiber simultaneously. Therefore, the modulation frequency of both waves and the delay from one side should be carefully set, so that only a single correlation peak along the fiber is measured at a time. Thus, the sensing range of a BOCDA system is usually short and limited by the distance between adjacent correlation peaks $d_m = v_g/(2f_m)$. The location of the measured correlation peak can be shifted by changing the modulation frequency.

In comparison with other Brillouin sensing techniques, BOCDA has an excellent performance in achieving high spatial resolution, which depends on the FM modulation amplitude and the natural Brillouin linewidth. Besides, since the acoustic wave is excited by CW waves, the BGS linewidth will not be broadened. However, since each point along the fiber must be measured individually, the total measurement time will be linearly proportional to the amount of resolved points, which makes the measurement time also longer, compared to the other Brillouin sensing techniques.

3.2.4 BOFDA

The principle of the Brillouin optical frequency domain analyzer (BOFDA) [49] is based on the measurement of the complex transfer function that relates the amplitudes of the CW counterpropagating pump and probe wave with the FUT. The probe wave is frequency downshifted to the pump by the BFS and amplitude modulated with a sinusoidal function at a variable frequency. The modulated pump and probe wave intensities are measured at the end of the FUT with two separate photodiodes (PD) that fed to a network analyzer (NWA). By sweeping the modulation frequency, the NWA measures the baseband transfer function of the FUT, whose inverse fast Fourier transformation approximates the pulse response of the FUT and represents the temperature and strain distribution. The high spatial resolution that BOFDA could achieve depends on the frequency sweep range, though, at the cost of the measurement time.

3.3 Basic setup of BOTDA

As illustrated in **Figure 6(a)**, a typical conventional BOTDA setup mainly comprises three parts. A highly coherent laser is split via a coupler into the probe (upper) and pump (lower) branch. A microwave signal, which can be scanned over the range of the BFS (~11 GHz ± 150 MHz), is applied on a Mach-Zehnder modulator (MZM 1) in the probe branch and biased in the carrier suppression regime. The generated lower frequency sideband serves as the probe wave, while the upper one is blocked by a narrow band-pass filter (FBG 1). The probe power is controlled by an erbium-doped fiber amplifier (EDFA 1) and launched into one end of the FUT. In the pump branch, the pump pulse with a required duration is formed by another MZM, amplified by EDFA 2 and launched into the other end of the fiber via a circulator (Cir 1). In order to mitigate the influence of the SBS polarization effect [50], the polarization of either pump or probe wave is scrambled via a polarization scrambler (Pol.S.) before launched into the fiber.

The Brillouin amplified probe wave is detected by a fast photodiode (PD) whose minimum bandwidth should correspond to the inverse of the pump pulse duration, to avoid any trace distortions. Another narrow band-pass filter (FBG 2) blocks the residual reflected pump wave. The electrical signal from the PD, which represents the evolution of the probe power, is processed digitally with a large number of averaging by a signal processor, that is, an oscilloscope or a digitizer. The Brillouin gain, which is essential for BFS estimation, can be derived by dividing the output of the PD by the probe power before the pump pulse is launched into the fiber [51].

3.4 Evaluation of the BOTDA performance

The BOTDA sensing performance is highly dependent on the estimation accuracy of the local BFS. According to recent investigations [52], system parameters such as the FWHM of the BGS $\Delta\nu_B$, the scanning frequency step δ, and the system noise σ, which is the inverse of the signal to noise ratio (SNR) of the normalized Brillouin gain, contribute significantly to an accurate estimation of a local BFS, as presented by the red curve in **Figure 6(b)**. Provided that $\nu_B(z)$ is the estimated BFS at distance z after a parabolic fitting of all experimental data above a given fraction η of the peak value (see **Figure 6(b)** for $\eta = 0.5$), the estimated error of the BFS is [52]:

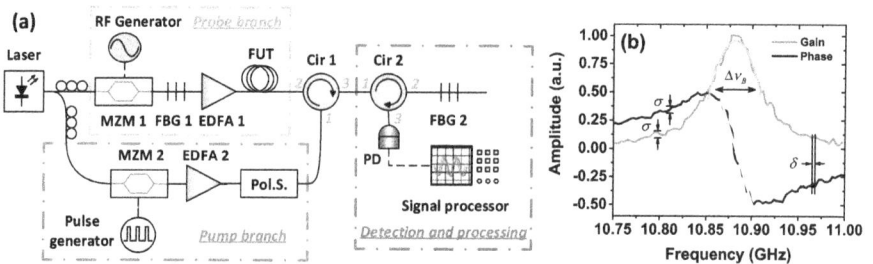

Figure 6.
(a) Conventional BOTDA setup. RF, radio frequency; MZM, Mach-Zehnder modulator; EDFA, erbium-doped fiber amplifier; Pol.S., polarization scrambler; FUT, fiber under test; PD, photodiode; Cir, circulator; FBG, fiber Bragg grating; (b) typical measured local BGS (red) and BPS (black) with system parameters that contribute to an uncertainty of the BFS estimation.

$$\sigma_V(z) = \sigma(z) \sqrt{\frac{3 \cdot \delta \cdot \Delta \nu_B}{8\sqrt{2}(1-\eta)^{3/2}}} \xrightarrow[\eta=0.5]{} \sigma(z) \sqrt{\frac{3}{4}\delta \cdot \Delta \nu_B} = \frac{1}{SNR(z)} \sqrt{\frac{3}{4}\delta \cdot \Delta \nu_B} \qquad (9)$$

As expected, a denser frequency sampling and a higher SNR value lead to a more accurate BFS estimation. Taking the relation of the SNR and number of averaging N_{AV} into consideration, that is, $SNR(z) \propto \sqrt{N_{AV}}$, the frequency error will also decrease significantly after thousands of averages due to an enhanced SNR. Owing to the linear dependence of the Brillouin phase response in the vicinity of the BFS, the sensor performance can also be evaluated by the linear fitting of the BPS [53] with a narrow frequency scanning and reduced measurement time (see black curve and yellow line in **Figure 6(b)**).

3.5 Major issues and limitations of BOTDA

3.5.1 Polarization fading

As discussed in Section 2.3, the Brillouin gain is highly dependent on the SOP of the pump and probe wave. Due to the weak birefringence of SMF, the SOP change of pump and probe leads to a highly nonuniform Brillouin gain along the fiber and consequently to a poor SNR for almost every fiber section, which is called polarization fading [50]. There are several solutions for polarization fading in a BOTDA system. The first proposed idea was to sequentially launch two orthogonal SOPs of the pump (or probe) wave and average the two measured results [4]. Another option is polarization scrambling, where the SOP of the pump and probe is varied rapidly so that the SOP is effectively randomized over time. It can be used in various scientific setups to cancel the errors caused by polarization-dependent effects. With a polarization scrambler and a digitizer, the traces can be averaged over thousands of pump pulses until a required high enough SNR is reached, though, at the cost of the BGS acquisition time.

3.5.2 Limitations on pump pulse width

According to the round-trip relation, the Brillouin interaction in a conventional BOTDA system takes place in a fiber section with a length of $cT/(2n)$, which can be seen as the spatial resolution [54], where c is the light speed in vacuum, n is the fiber refractive index and, T is the pulse duration. In principle, the spatial resolution could be enhanced by using shorter pulses. However, two main factors limit the spatial resolution to only 1 m.

First of all, decreasing the pulse duration will shorten the Brillouin interaction length and hence lower the SBS gain and consequently the SNR. Furthermore, the pump power spectrum for a shorter pulse will be severely broadened. Therefore, the resulting effective BGS should be modified as the convolution of the pump spectrum density and the natural Brillouin linewidth [21], that is,

$$g_B^{eff}(\nu) = g_B(\nu) \otimes I_p(\nu) \qquad (10)$$

where $I_p(\nu)$ is the pump pulse power spectrum, which is significantly broadened to $1/T$ for pulses shorter than 20 ns [55]. The inherent broadening of the BGS leads to a decrease of the peak gain, which makes the estimation of the BGS peak more sensitive to the noise level. Moreover, it also indicates that the probe wave will be amplified only after the acoustic field is excited by the pump-probe interaction, which takes 10–30 ns. Due to the abovementioned reasons, the spatial resolution of a conventional BOTDA system is limited to around 1 m, which corresponds to a pulse width of 10 ns.

3.5.3 Limitations on the pump wave power

In principle, the pump pulse power launched into the FUT should be high enough to compensate the fiber attenuation and hence generate an efficient Brillouin interaction. However, modulation instability (MI) in the optical fiber appears when the pulse power is beyond a certain threshold [56]. MI refers to the breakup of the balance between the anomalous dispersion and the Kerr effect, so that a train of soliton-like pulses rises from the noise as spectral sidebands symmetric to the pulse frequency [57, 58]. Several theoretical as well as experimental investigations have also demonstrated that there exists a periodical energy exchange between sidebands and pulse power along the fiber, that is, after a certain length of propagation, the pulse energy that has been spread to the sidebands will be transferred back to the pulse frequency and thus forms the Fermi-Pasta-Ulam recurrence [59].

For a conventional BOTDA system, this periodical power exchange leads to the fluctuation of the pump pulse power during the propagation and therefore distortions of the traces. As illustrated in **Figure 7(a)**, some of the fiber sections are not correctly interrogated due to the limited SNR. Furthermore, the critical power of the MI for a conventional BOTDA system with 25 km fiber is estimated to be as low as 135 mW [57]. Due to a rapid decrease of the Brillouin gain, MI severely limits the sensing range.

3.5.4. Limitations on the probe wave power

In order to achieve a high SNR for a better BFS estimation, a high enough probe wave power should be launched into the FUT. However, due to nonlocal effects (NLEs), the probe wave power in a conventional BOTDA system, that is, single probe sideband system, is usually limited to only −14 dBm [60]. The NLE generally refers to the fact that the Brillouin interaction from a local fiber segment is influenced by the interaction of other segments and hence leads to an error in the BFS estimation of the local fiber segment. It should be noticed that not only a too high probe power, but other factors, such as a limited pump pulse extinction ratio (ER), would also lead to NLE.

The origin of the NLE due to the high probe wave can well be explained by **Figure 7(b)** and the DES of SBS. As pointed out by Eq. (2) in Section 2.1, the

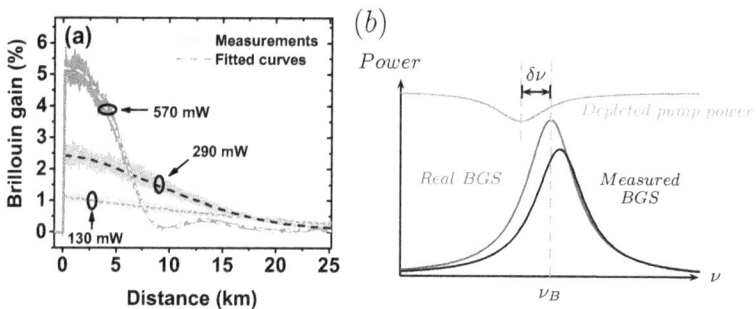

Figure 7.
(a) Distorted Brillouin traces due to MI with different input pump powers and fitted curves [57]. (b) Schematic explanation of the NLE due to the pump depletion in the FUT with uniform BFS in a long distance but nonuniform BFS in a distant section. $\delta\nu$ is the frequency shift of BGS between the long uniform sections and the distant fiber segment [60].

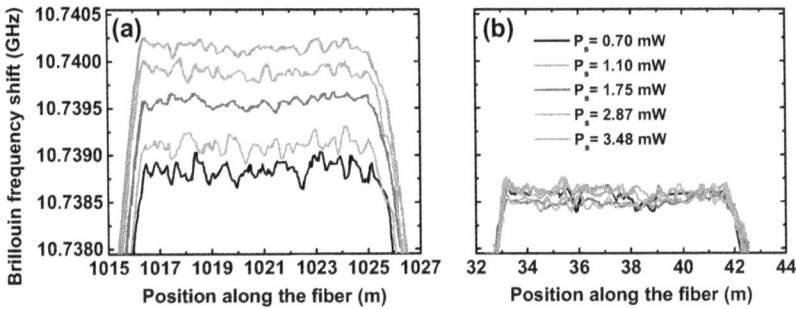

Figure 8.
BFS as a function of the position along the fiber when the hot spot is placed at (a) the far end and (b) the near end of the fiber for different signal (probe) powers P_s [60].

Brillouin interaction at each fiber segment not only amplifies the probe wave but also slightly depletes the pump power intensity. Since for each time-resolved interval, the CW probe wave interacts with the pump pulse only once, and the impact of the depletion becomes much higher on the pulse than on the probe wave. Therefore, the pump spectrum density is no longer independent of the Brillouin interaction that the pump pulse has experienced before reaching the distant fiber segment with nonuniform BFS (see **Figure 7(b)**). NLE typically leads to an asymmetry of the BGS and hence to an error in the BFS estimation. It should be noticed that, due to the accumulative impact of the depletion on the pump pulse, the NLE is in general more severe at the far end of the fiber (see **Figure 8**).

Another limitation on the probe power is the Brillouin threshold of the fiber. It is theoretically the maximum power that the input probe wave can have with no depletion and no thermally induced SpBS from the probe wave [61]. Since the Brillouin threshold is usually higher than the probe power limit due to the NLE, a series of techniques have to be applied to solve the NLE and then push the probe power limit toward the threshold of SBS.

4. State-of-the-art of BOTDA

As discussed in Section 3, due to the limitations on the pump pulse and probe power, the spatial resolution, and hence the sensing range and BFS measurement accuracy, a conventional BOTDA system is far from achieving an ideal performance. In order to compensate the SNR degradation, a higher averaging must be applied at the cost of measurement time. In this section, methods to break these limitations are reviewed.

The techniques introduced in this section are categorized according to their enhanced sensor performances. It is worth to notice that the contribution of each technique may lead to enhancements in several performances, for example, a technique that overcomes the MI enables a pump power higher than the limit and thus also an extended sensing range due to the increased Brillouin gain.

4.1 Strategies to avoid modulation instability

4.1.1 Noise filtering

The origin of MI is system noise, in which especially the amplified spontaneous emission (ASE) noise from the EDFA for pump pulse amplification plays an

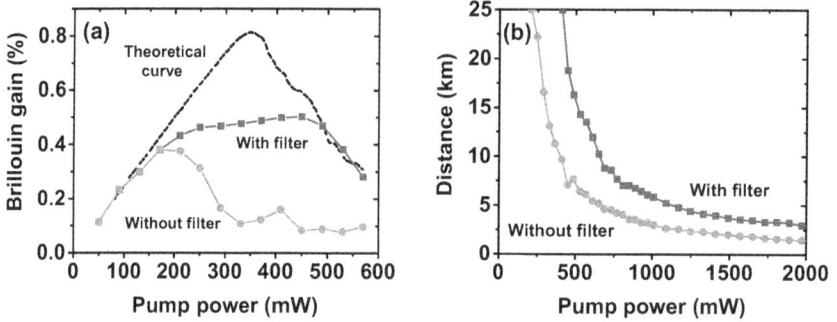

Figure 9.
Experimental results of the (a) Brillouin gain and (b) maximum distance (b) with an increasing input pump power with and without ASE noise filtering [62].

important factor. Therefore, a narrow band-pass optical filter after the pulse amplification might mitigate the MI [62]. Conventionally, the Brillouin gain is proportional to the input pump power, as indicated in **Figure 9(a)**. However, this is not the case if MI takes place. When the pump is depleted, the Brillouin gain decreases correspondingly. As indicated by the red line in **Figure 9(a)**, the application of the filter enables a slight increase of the Brillouin gain when the pump power is beyond the MI threshold, indicating a lower pump depletion in comparison to the case without the 1 GHz band-pass filter, and a mitigation of MI. Besides, the sensing range is also extended due to the lower pump depletion. **Figure 9(b)** illustrates experimental results for the distance that the pump pulse reaches when its amplitude is depleted by 50%. Since the MI extends over a band of ~60 GHz [62], it is believed that a filter wider than this bandwidth has a negligible contribution on the MI mitigation, while an enhanced mitigation is expected with a much narrower filter.

4.1.2 Dispersion-shifted fiber

Since the origin of MI is the interplay between the Kerr effect and anomalous dispersion, the MI can be eliminated in a dispersion-shifted fiber (DSF) with normal dispersion $D = -1.4$ ps/(nm·km) at 1550 nm [63]. As discussed in Section 3.5.3, the MI threshold in a 25 km SMF is estimated to be around 135 mW. However, as shown in **Figure 10(a)**, in DSF with the same length, no obvious trace distortion is observed at a pump pulse power of 400 mW, indicating a much higher MI threshold. An exponential fitting, which implies a pure fiber attenuation, further confirms the full

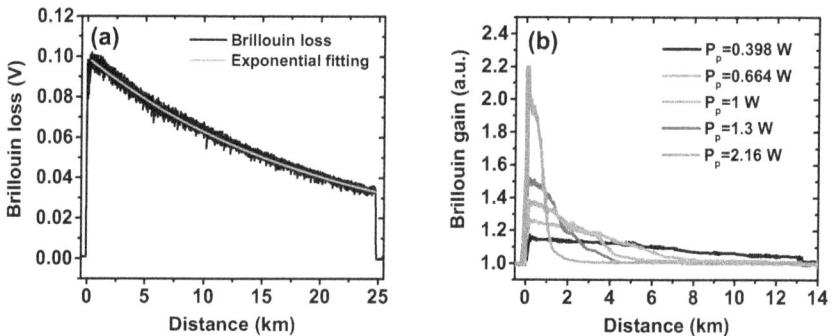

Figure 10.
(a) BOTDA trace in 25 km DSF with 400 mW pump power [63]; (b) distortions due to forward Raman scattering [64].

mitigation of MI. In comparison with other techniques, this technique is simple and requires no further change in the setup. However, DSFs of long length are much more costly than SMFs of the same length and DSFs usually have a higher attenuation.

Though the successful mitigation, according to a recent investigation [64], there are still pump power limitations. As shown in **Figure 10(b)**, for a pump pulse with a power of around 1 W, the traces are distorted mainly by forward Raman scattering.

4.1.3 Orthogonal polarized pump pulses

Since the single pump pulse power is limited by MI, a novel technique based on multiwavelength pumps was proposed to break the power limitation [65]. However, it has soon been identified that four wave mixing (FWM) between different spectral lines [65, 66] limits this approach. Thus, an orthogonal polarized dual-pump technique was proposed [67]. In this technique, two pump pulses with orthogonal SOP in different frequencies are launched into the FUT and interact with their corresponding probe waves. As shown in **Figure 11**, the orthogonal pumps effectively mitigate the MI in comparison with a single pump regime and successfully avoid the FWM interaction in comparison with the parallel polarized dual-pump regime both in time and frequency domain. It is worth to note that no additional polarization scrambler is required for orthogonal pulses, since two complementary Brillouin interactions are already ensured.

4.2 Strategies to avoid nonlocal effects

In an early proposal to avoid NLE, a general postprocessing algorithm was used for the BFS profile reconstruction [45]. Based on the BFS distribution that matches the measured data with a minimization algorithm in multidimensions, the parameters of the unknown BFS profile can be derived. However, the experimental realization complicates the setup and increases the processing time [68]. Another idea for the mitigation of NLE in general is to use pulses for both pump and probe wave [69]. However, despite the effective mitigation of the NLE, the shortened interaction length increases the measurement time significantly. In this subsection, other effective solutions for the NLE are reviewed.

4.2.1 Dual probe

One of the origins of NLE is the pump pulse depletion, which can be further separated in first- and second-order NLE. First-order NLEs are mainly caused by

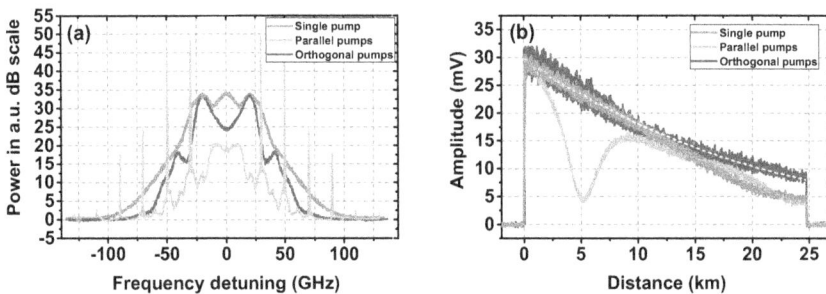

Figure 11.
(a) Simulated MI gain spectrum and (b) experimental time traces with single pump and parallel/orthogonal polarized dual-pump regime with ±10 GHz frequency detuning [67].

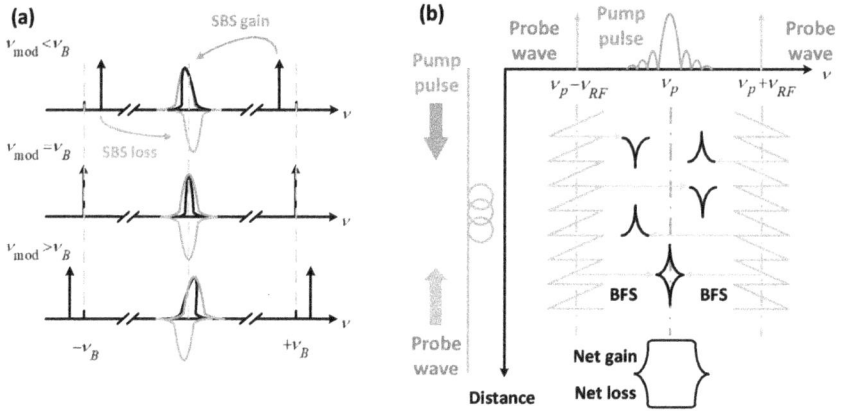

Figure 12.
Schematic explanation of (a) the Second-order NLE due to the distortion of the pump spectrum [70] and (b) the strategy to avoid it using an FM probe wave [73].

pump depletion due to a high probe power. Second-order NLEs are based on linear distortions of the pump pulse spectrum. Second-order NLEs can even happen when first-order NLEs are completely mitigated [70].

The most popular solution for first-order NLE mitigation is the dual-probe sideband regime. In contrast to a conventional BOTDA system, both probe wave sidebands generated by the MZM are involved in the SBS interaction. The pump pulse depletion due to the energy transfer to the lower frequency sideband is compensated by an energy transfer from the upper frequency sideband to the pump. A theoretical analysis reveals that in the dual-probe sideband regime, the probe power limit rises from -14 to -3 dBm [60, 71]. Furthermore, since the energy transfer from the high power (pump) to the low power (probe) is more efficient than the reverse process, slightly unbalanced dual-probe sidebands miti-gate NLE better than balanced ones [60, 72].

Even with dual probe regime, second-order NLEs still exist [70]. This kind of NLE has its origin not in the depleted pump, but in the frequency-dependent distortion of the pump pulse spectrum that affects the interaction in the gain and loss configuration differently (see **Figure 12(a)**). However, a total mitigation of first- and second-order NLEs can be achieved with dual-probe sidebands [61] with additional FM in saw-tooth shape [73]. A similar performance enhancement can also be achieved with sinusoidal or a triangular shape [61, 74]. If the FM is syn-chronized to the pump pulses (see **Figure 12(b)**), a series of pulses at a specific fiber position interacts always with the same probe frequency. In turn, each single pulse experiences SBS interactions with probe waves that have different frequency detunings as it propagates along the fiber. Since the pulse depletion is accumulated along the fiber, a decreased distortion upon the pulse spectrum can be achieved by interacting with different frequencies within the BGS, indicating a mitigation of the second-order NLE. Together with the mitigation of the first-order NLE by dual-probe sidebands regime, the probe power limit has been successfully pushed to a record of 8 dBm [61], reaching the Brillouin threshold of SMF in $\sim$20 km range.

4.2.2 Higher pulse extinction ratio

Another origin of NLE is the imperfect pulsing of the pump wave. Due to the limited extinction ratio (ER) of common pulsing methods, there is a residual power leakage. Since the pulse pedestals have the same frequency as the pump, SBS

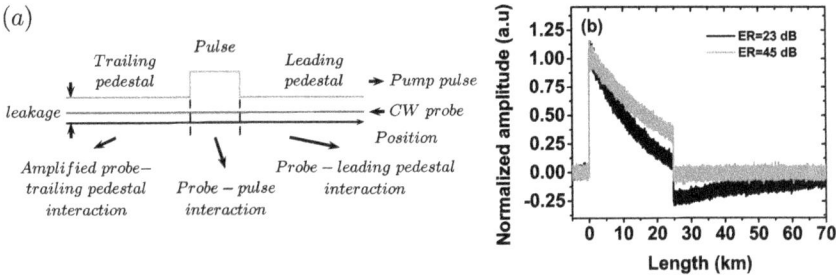

Figure 13.
(a) The schematic description of the interaction between pulse and probe wave [75]; (b) the distorted BOTDA traces under different pulse ER due to the pump pedestal depletion [76].

interactions can take place between the probe wave and the leading as well as the trailing pedestals [75], as schematically depicted in **Figure 13(a)**. Therefore, the net effect of the presence of the pulse pedestal is that the probe wave will be additionally amplified. However, this amplification is not useful for sensing. In case of a strong probe wave, not only the pulse but also the pedestals (especially the trailing pedestal) will deplete, which lead to severe distortions on the BOTDA traces (see **Figure 13(b)**) [75, 76].

Under a very low ER, the interactions between probe wave and pedestals may dominate the final BGS detection due to a long interaction length and leads to an error on the BFS estimation (see **Figure 14(a)** and **(b)**) [77]. For a conventional MZM, the ER is only 20 dB but can be enhanced up to 30 dB with special designs. Higher ERs of more than 40 dB can be achieved with switching type semiconductor optical amplifiers (SOA) and a 60 dB ER with RF switches [75, 76].

4.3 Enhancement of the spatial resolution

4.3.1 Differential pulse width pair (DPP)

In a DPP-BOTDA system, two consecutive measurements are carried out with the same pulse peak power but slightly different durations, T and $T + \Delta T$. Usually, T is longer than the phonon excitation time, indicating a full excitation of the acoustic wave in both measurements and ΔT is short so as to achieve a high spatial resolution [78]. Since both pump pulses give the same amplification in the time slot T and only the longer pulse contributes amplification in the extra time ΔT, the subtraction of the two measured Brillouin amplified probe signals yields to the BGS in the time slot ΔT.

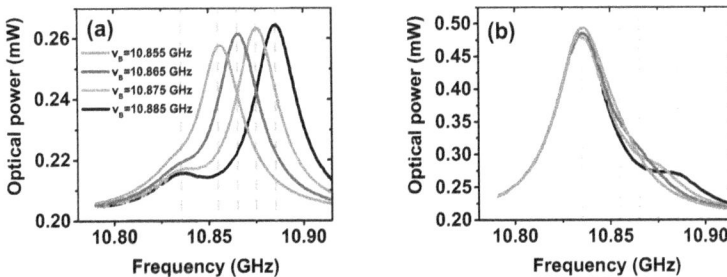

Figure 14.
The simulation of the BGS with pump pulse of (a) ER = 40 dB and (b) ER = 26 dB at the hot spots in different temperatures (different BFSs). The uniform BFS of the rest of the fiber is $\nu_{B0} = 10.835$ GHz [77].

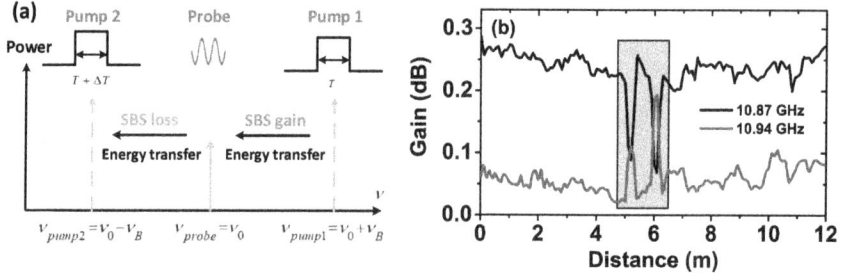

Figure 15.
(a) Schematic description of the simultaneous DPP-BOTDA in the frequency domain and (b) Brillouin gain profile in the experiment. Along the fifth meter of the fiber, two 20 cm fiber sections were located 65 cm apart and manually stretched to have nonuniform strains [79].

Later on, the DPP-BOTDA technique has been developed into a simultaneous measurement by launching two pump pulses with slightly different durations in different frequencies [79]. Pump pulse 1 with duration T interacts with the probe wave via SBS loss, while simultaneously pump pulse 2 with duration $T + \Delta T$ interacts with the probe wave via SBS gain (see **Figure 15(a)**). Therefore the subtraction of the BGS is automatically achieved at the detector with no postprocessing required. A 10 cm spatial resolution BOTDA system has been reported by using a 30 ns gain pump pulse and a 29 ns loss pump pulse (see **Figure 15(b)**) [79]. However, in comparison with the pre-excitation method, the subtraction of the traces adds noise to the data. Therefore, in order to achieve the required SNR, massive averaging must be applied.

4.3.2 Pre-excitation

The reason for the spatial resolution limit of around 1 m is the excitation time of the phonon. A prepump excitation can solve this problem by shaping the pump pulse into two parts, that is, a long pedestal with low power (prepump pulse (PPP), part 1 in **Figure 16(a)**) for the phonon excitation, followed by a narrow high power pulse (part 2 in **Figure 16(a)**) [80]. In order to excite the phonon, the PPP is usually longer than 10 ns. To achieve high spatial resolutions, the high power pulse can be very short (~1 ns). **Figure 16(b)** shows experimental results when the PPP (12 ns duration), the high power pulse (1 ns duration), and the total pump pulse interrogate a 20 cm fiber section with strain individually. The resulting BGS of the total

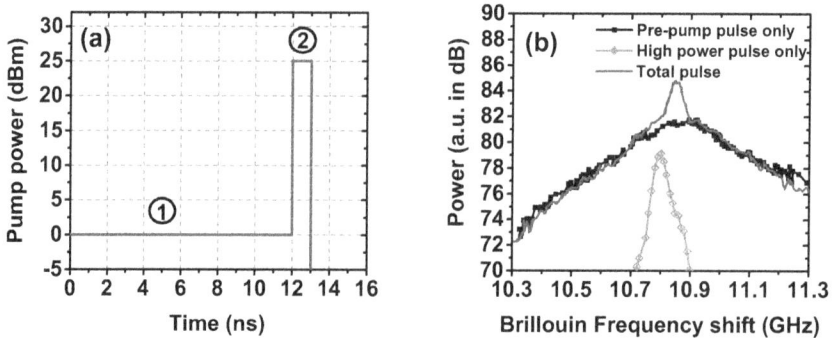

Figure 16.
(a) Pulse shape for pre-excitation technique; (b) BGS of a fiber section with strain interrogated by the PPP only (red), high power pulse only (black), and total pulse (blue) [80].

pulse is characterized by a broadened spectrum with low amplitude and a narrow cap, whose width is determined by the high power pulse and PPP, respectively. A clear BFS shift between the PPP and total pulse case indicates an enhancement of the spatial resolution down to ~10 cm. In comparison to DPP-BOTDA, the high spatial resolution by pre-excitation does not come at the expense of a decreased SNR and therefore, it is more favorable for commercial use.

4.4 Enhancement of the sensing range

Due to the fiber attenuation or other depletions, the sensing range of a conventional BOTDA system is usually limited by the low SNR at the far end of the fiber to only a few tens of kilometers [51]. Therefore, the key to extend the sensing range is either to enhance the probe signal power or to eliminate the system noise.

4.4.1 Multi-frequency pump-probe interaction

In this technique, the total pump power is spread over multiple-pump waves in different frequencies, with every single pump power still limited by MI [65]. The theoretical enhancement of the SNR could reach the number of pumps N. However, severe FWM occurs for too narrow pump frequency spacing, while the BGS linewidth from each pump may differ when they are too widely separated apart [65]. The solution for the latter is a postprocessing algorithm [81], while the solution to avoid the FWM is to shift the pump pulse propagation in time domain with a frequency-selective time shifter, which can be realized by N-consecutive FBGs separated by a certain length of fiber in the experiment. The schematic description of the frequency-selective time shifter is illustrated in **Figure 17(a)**. After the time-shifted pump pulses have interacted with their corresponding probe waves, another consecutive FBG with a reversed sequence offers a reversed delay and combines the traces back in time domain so that they can be simultaneously detected. For a three pump system, an SNR improvement of 4.8 dB has been demonstrated (see **Figure 17(b)**) [65].

4.4.2 Self-heterodyne detection

Another possibility to amplify the signal amplitude is the heterodyne detection. Provided that the Brillouin amplified probe wave at frequency ν_s beats with an local oscillator at frequency ν_{LO}, the total electrical field can be expressed as [30]:

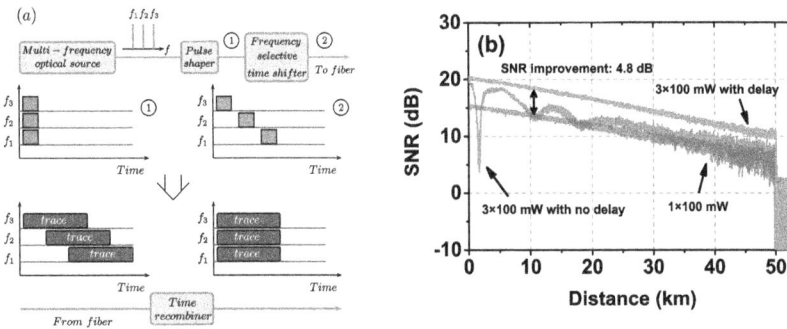

Figure 17.
(a) Schematic explanation of the time shifter and recombiner; (b) SNR measured in the experiment for standard BOTDA (single pulse), three pulses with and without time delay [65].

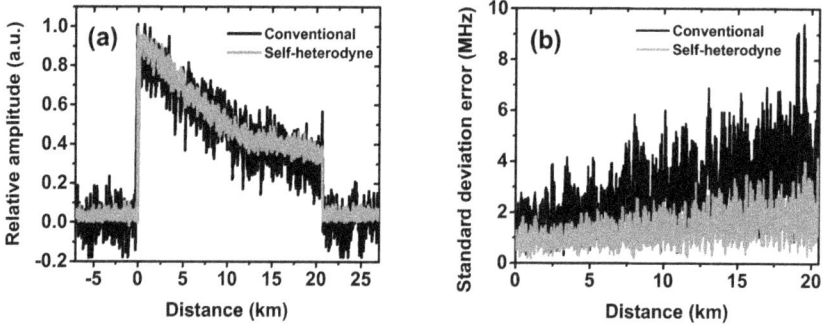

Figure 18.
(a) BOTDA traces and (b) BFS standard deviation of consecutive measurements with (red) and without (black) self-heterodyne detection [30].

$$E_T(t,\nu) = E_{S0}\, g_{SBS}(\nu_s,z)\, \exp\left\{j \cdot [2\pi\nu_s t + \phi_{SBS}(\nu_s,z)]\right\} + E_{LO}\, \exp\left(j \cdot 2\pi\nu_{LO}t\right) \quad (11)$$

where E_{s0} and E_{LO} are the complex amplitude of the probe wave and local oscillator, g_{SBS} and ϕ_{SBS} are the SBS gain and phase shift. Hence, the detected current at the PD can be written as:

$$I_c(t) = R_c P_T = 2R_C \sqrt{P_{s0}\left[1 + g_{SBS}(\nu_s,z)\right]^2 P_{LO}} \cdot \cos\left[2\pi f_{IF} t + \phi_0 - \phi_{SBS}(\nu_s,z)\right]$$

$$(12)$$

where ϕ_0 and f_{IF} are the phase and frequency difference between the probe wave and local oscillator, P_{LO} and P_{S0} are the optical power of the local oscillator and probe wave, R_c is the PD responsivity. As the detected current is dependent on the power of the local oscillator, the whole mechanism could be regarded as a signal amplification with a strong oscillator. As shown in **Figure 18(a)**, the trace detected by self-heterodyne is cleaner and has an SNR enhancement of 10 dB [30]. Further investigations with five consecutive distributed measurements show that the self-heterodyne detection decreases also the BFS standard deviation (see **Figure 18(b)**), indicating a more accurate BFS estimation. In comparison with other techniques, self-heterodyne detection provides the most simple and feasible scheme for sensing range enhancement.

4.5 Enhancement on measurement time

For a conventional BOTDA, it usually takes several minutes to finish a single measurement, which is impractical for dynamic strain sensing. One of the main factors that limit the measurement time is the sweeping of the probe frequency to scan the total BGS. Therefore, several techniques have been applied to solve this problem.

4.5.1 Slope-assisted BOTDA

One of the techniques for dynamic sensing is based on partially scanning the BGS [82]. In order to achieve the BGS profile in general, the technique requires a preliminary frequency scan without vibration. The probe scanning frequency is then set at half of the BGS linewidth, as shown in **Figure 19(a)**. This 3 dB point has the steepest slope and widest linear range and is the most sensitive working point for tiny frequency shifts. The BGS is reconstructed according to the measured signal amplitude, and the strain values are obtained by the strain coefficient C_ε mentioned

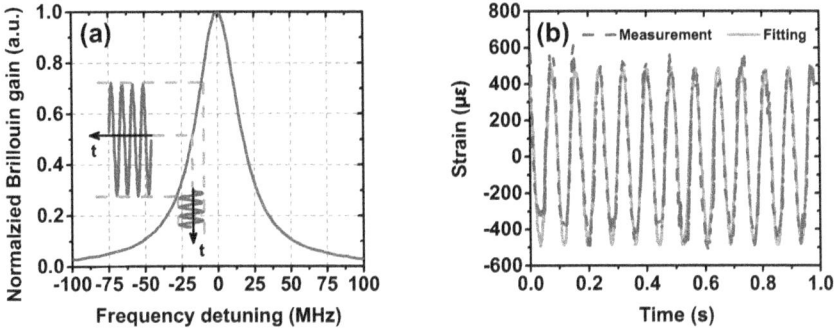

Figure 19.
(a) Schematic explanation of the dynamic sensing with the working point set at the half value of the peak Brillouin gain and (b) dynamic strain measurement based on this technique [82].

in Section 3.1. In **Figure 19(b)**, the proposed technique is demonstrated with a perturbation frequency of 12.3 Hz. Obviously, this technique can only detect small temperature or strain distributions and is not applicable for a BFS nonuniformity that exceeds half of the BGS linewidth.

4.5.2 Sweep-free multi-tones

Another idea to speed up the measurement time is to utilize multitone pumps [66]. As shown in **Figure 20(a)**, multiple pumps with multiple frequencies are launched into the fiber aligning at different frequency detunings of the BGSs. Since the multi-SBS interactions happen simultaneously, the necessary frequency sweeping can be done in a single shot. Thus, the total measurement time including extra averaging measurements for SNR improvement lasts only several seconds.

However, there are still trade-offs and limitations in this technique. Undesired FWM occurs when the pump pulses are simultaneously launched. This can be solved by unequal spacing or a sequential launch of the pumps. The total number of pump tones, which determines the accuracy of the BGS reconstruction, is restricted by the intertone spacing Δf and the total tone span f_{total}. The intertone spacing is usually larger than the total BGS width so as to avoid BGS overlaps. The total tone span should not exceed the BFS so that no pump lines are within the BGS of other pumps.

Recently, a new improved method has been proposed to avoid these limitations by utilizing digital optical frequency combs (DOFC) as probe signals [83]. Since the probe wave power in the BOTDA system is usually low, multiple-probe waves suffer less from FWM than multiple pumps. The DOFC owns narrow frequency spacing, wide flat top, and total bandwidth, as depicted in **Figure 20(b)**. According

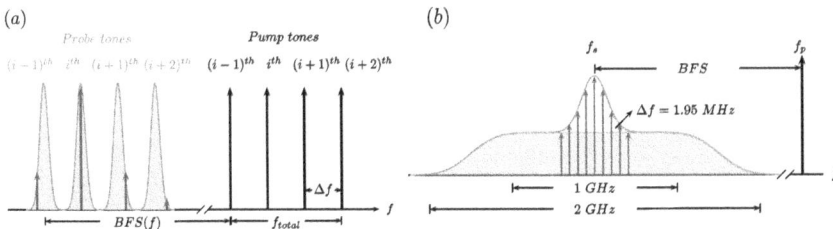

Figure 20.
(a) Schematic explanation of the principle of the sweep-free multitone BOTDA [81]; (b) spectrum of DOFC [83].

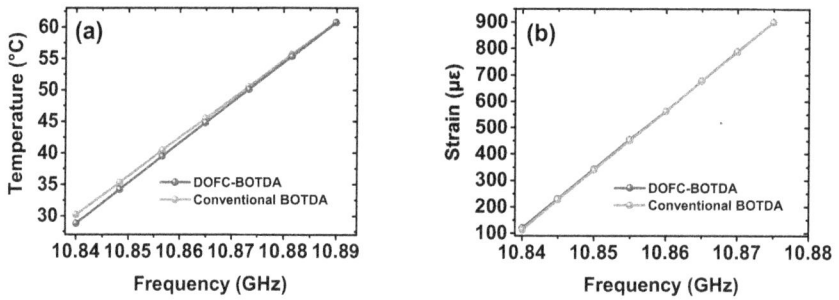

Figure 21.
The linear fitting of (a) temperature and (b) strain measurement with conventional BOTDA and DOFC-BOTDA [83].

to the frequency difference between each line and the pump frequency, the pump pulse shapes the amplitude of each spectral comb line via the SBS interaction simultaneously. The total measurement time, including the necessary 100 acquisitions for the SNR improvement, results in only 10 ms [83]. Furthermore, the experimental results in **Figure 21** confirm its equivalence to a conventional BOTDA regarding the temperature and strain measurement. However, the main disadvantage of this technique is the special requirement of the DOFC, which is generally not so simple to achieve.

5. Summary

In this chapter, the basics of SBS and its application for distributed sensing have been reviewed. The overview has started with an introduction of SBS together with its physical origin and applications due to inherent, striking advantages in a variety of fields such as slow light, optical and microwave photonic filters, and many more. Among all these exciting applications, distributed temperature and strain sensing is one of the most prominent.

The enhanced SNR and the moderate resolution are the superiority of distributed Brillouin sensors to the traditional distributed and point sensors in long-range sensing. However, conventional BOTDA sensors are limited by MI and NLE. The origins, as well as methods for the mitigation of MI and NLE, have been presented and discussed in detail. Thus, with these new methods, much longer sensing ranges became possible.

Besides the sensing range, methods to enhance the spatial resolution and the speed of the measurement have also been reviewed and discussed. Nowadays, distributed Brillouin fiber sensors can have a resolution in the centimeter range, or even below and act like thousands or millions of point sensors. At the same time, novel ideas such as multi-tone pumps have successfully shortened the measurement time in distributed SBS sensors from several minutes down to ~10 ms. Due to the fruitful proof-of-concept results, some of the state-of-the-art techniques discussed in this chapter have already been applied in some BOTDA prototypes.

Acknowledgements

Cheng Feng wishes to acknowledge the financial support from German Research Foundation (DFG SCHN 716/13-1) and Niedersächsisches Vorab (NL—4 Project

"QUANOMET"). Jaffar Emad Kadum would like to acknowledge the financial support of Iraqi Ministry of Oil/State Company for Oil Projects (SCOP).

Author details

Cheng Feng*, Jaffar Emad Kadum and Thomas Schneider
Institute for High Frequency Technology, Technical University of Braunschweig, Braunschweig, Germany

*Address all correspondence to: cheng.feng@ihf.tu-bs.de

IntechOpen

© 2019 The Author(s). Licensee IntechOpen. This chapter is distributed under the terms of the Creative Commons Attribution License (http://creativecommons.org/licenses/by/3.0), which permits unrestricted use, distribution, and reproduction in any medium, provided the original work is properly cited.

References

[1] López-Higuera JM, editor. Handbook of Optical Fibre Sensing Technology. Chichester: Wiley; 2002

[2] Barnoski MK, Rourke MD, Jensen SM, Melville RT. Optical time domain reflectometer. Applied Optics. 1977;**16**: 2375-2379. DOI: 10.1364/AO.16.002375

[3] Grattan KTV, Sun T. Fiber optic sensor technology: An overview. Sensors and Actuators A: Physical. 2000;**82**:40-61. DOI: 10.1016/S0924-4247(99)00368-4

[4] Horiguchi T, Tateda M. Optical-fiber-attenuation investigation using stimulated Brillouin scattering between a pulse and a continuous wave. Optics Letters. 1989;**14**:408-410. DOI: 10.1364/OL.14.000408

[5] Brillouin L. Diffusion de la lumière et des rayons X par un corps transparent homogène. Annals of Physics. 1922;**9**: 88-122. DOI: 10.1051/anphys/192209170088

[6] Ippen EP, Stolen RH. Stimulated Brillouin scattering in optical fibers. Applied Physics Letters. 1972;**21**: 539-541. DOI: 10.1063/1.1654249

[7] Agrawal GP. Nonlinear Fiber Optics. 5th ed. Amsterdam: Elsevier, Academic Press; 2013

[8] Shelby RM, Levenson MD, Bayer PW. Guided acoustic-wave Brillouin scattering. Physical Review B. 1985;**31**: 5244-5252. DOI: 10.1103/PhysRevB.31.5244

[9] Antman Y, London Y, Zadok A. Scanning-free characterization of temperature dependence of forward stimulated Brillouin scattering resonances. In: Kalinowski HJ, Fabris JL, Bock WJ, editors. 24th Int. Conf. Opt. Fibre Sensors. Vol. 9634. 2015. p. 96345C. DOI: 10.1117/12.2195097

[10] Antman Y, Clain A, London Y, Zadok A. Optomechanical sensing of liquids outside standard fibers using forward stimulated Brillouin scattering. Optica. 2016;**3**:510. DOI: 10.1364/OPTICA.3.000510

[11] Boyd RW. Nonlinear Optics. 3rd ed. Singapore: Elsevier; 2008

[12] Smith RG. Optical power handling capacity of low loss optical fibers as determined by stimulated Raman and Brillouin scattering. Applied Optics. 1972; **11**:2489. DOI: 10.1364/AO.11.002489

[13] Küng A. Laser Emission in Stimulated Brillouin Scattering in Optical Fiber. Lausanne, Switzerland: Ecole Polytechnique Fédérale de Lausanne; 1997

[14] Bayvel P, Radmore PM. Solutions of the SBS equations in single mode optical fibres and implications for fibre transmission systems. Electronics Letters. 1990;**26**:434. DOI: 10.1049/el:19900282

[15] Le Floch S, Cambon P. Theoretical evaluation of the Brillouin threshold and the steady-state Brillouin equations in standard single-mode optical fibers. Journal of the Optical Society of America. A. 2003;**20**:1132. DOI: 10.1364/JOSAA.20.001132

[16] Engelbrecht R. Analysis of SBS gain shaping and threshold increase by arbitrary strain distributions. Journal of Lightwave Technology. 2014;**32**: 1689-1700. DOI: 10.1109/JLT.2014.2310214

[17] Zadok A, Eyal A, Tur M. Stimulated Brillouin scattering slow light in optical fibers [invited]. Applied Optics. 2011; **50**:E38. DOI: 10.1364/AO.50.000E38

[18] Wei W, Yi L, Jaouën Y, Hu W. Bandwidth-tunable narrowband rectangular optical filter based on

stimulated Brillouin scattering in optical fiber. Optics Express. 2014;**22**: 23249-23260. DOI: 10.1364/OE.22.023249

[19] Feng C, Preussler S, Schneider T. Sharp tunable and additional noise-free optical filter based on Brillouin losses. Photonics Research. 2018;**6**:132-137. DOI: 10.1364/PRJ.6.000132

[20] Stern Y, Zhong K, Schneider T, Zhang R, Ben-Ezra Y, Tur M, et al. Tunable sharp and highly selective microwave-photonic band-pass filters based on stimulated Brillouin scattering. Photonics Research. 2014;**2**:B18-B25. DOI: 10.1364/PRJ.2.000B18

[21] Zhu Z, Dawes AM, Gauthier DJ, Zhang L, Willner AE. 12-GHz-bandwidth SBS slow light in optical fibers. In: Opt. Fiber Commun. Conf.; Optical Society of America. 2006. p. PDP1

[22] Schneider T, Junker M, Lauterbach K-U. Potential ultra wide slow-light bandwidth enhancement. Optics Express. 2006;**14**:11082-11087. DOI: 10.1364/OE.14.011082

[23] Schneider T, Henker R, Lauterbach K-U, Junker M. Adapting Brillouin spectrum for slow light delays. Electronics Letters. 2007;**43**:682. DOI: 10.1049/el:20070313

[24] Preussler S, Wiatrek A, Jamshidi K, Schneider T. Brillouin scattering gain bandwidth reduction down to 3.4 MHz. Optics Express. 2011;**19**:8565-8570. DOI: 10.1364/OE.19.008565

[25] Wiatrek A, Preußler S, Jamshidi K, Schneider T. Frequency domain aperture for the gain bandwidth reduction of stimulated Brillouin scattering. Optics Letters. 2012;**37**: 930-932. DOI: 10.1364/OL.37.000930

[26] Preussler S, Schneider T. Bandwidth reduction in a multistage Brillouin system. Optics Letters. 2012;**37**: 4122-4124. DOI: 10.1364/OL.37.004122

[27] Preussler S, Schneider T. Stimulated Brillouin scattering gain bandwidth reduction and applications in microwave photonics and optical signal processing. Optical Engineering. 2016;**55**:031110. DOI: 10.1117/1.OE.55.3.031110

[28] Preussler S, Wiatrek A, Jamshidi K, Schneider T. Quasi-light-storage enhancement by reducing the Brillouin gain bandwidth. Applied Optics. 2011;**50**: 4252-4256. DOI: 10.1364/AO.50.004252

[29] Feng C, Preussler S, Schneider T. The influence of dispersion on stimulated Brillouin scattering based microwave photonic notch filters. Journal of Lightwave Technology. 2018; **36**:5145-5151. DOI: 10.1109/JLT.2018. 2871037

[30] Zornoza A, Sagues M, Loayssa A. Self-heterodyne detection for SNR improvement and distributed phase-shift measurements in BOTDA. Journal of Lightwave Technology. 2012;**30**: 1066-1072. DOI: 10.1109/JLT.2011. 2168808

[31] Horiguchi T, Shibata N, Azuma Y, Tateda M. Brillouin gain variation due to a polarization-state change of the pump or stokes fields in standard single-mode fibers. Optics Letters. 1989;**14**:329. DOI: 10.1364/OL.14.000329

[32] Sharma GP, Preubler S, Schneider T. Precise optical frequency shifting using stimulated Brillouin scattering in optical Fibers. IEEE Photonics Technology Letters. 2017;**29**:1467-1470. DOI: 10.1109/LPT.2017.2729598

[33] Al-Taiy H, Wenzel N, Preußler S, Klinger J, Schneider T. Ultra-narrow linewidth, stable and tunable laser source for optical communication systems and spectroscopy. Optics Letters. 2014;**39**:5826-5829. DOI: 10.1364/OL.39.005826

[34] Preussler S, Zadok A, Wiatrek A, Tur M, Schneider T. Enhancement of spectral resolution and optical rejection ratio of Brillouin optical spectral analysis using polarization pulling. Optics Express. 2012;**20**:14734-14745. DOI: 10.1364/OE.20.014734

[35] Preussler S, Schneider T. Attometer resolution spectral analysis based on polarization pulling assisted Brillouin scattering merged with heterodyne detection. Optics Express. 2015;**23**: 26879-26887. DOI: 10.1364/OE.23.026879

[36] Preussler S, Wenzel N, Schneider T. Flat, rectangular frequency comb generation with tunable bandwidth and frequency spacing. Optics Letters. 2014;**39**:1637-1640. DOI: 10.1364/OL.39.001637

[37] Preußler S, Wenzel N, Braun R-P, Owschimikow N, Vogel C, Deninger A, et al. Generation of ultra-narrow, stable and tunable millimeter- and terahertz-waves with very low phase noise. Optics Express. 2013;**21**:23950-23962. DOI: 10.1364/OE.21.023950

[38] Preussler S, Wenzel N, Zadok A, Schneider T. Tunable generation of ultra-narrow linewidth millimeter and THz-waves and their modulation at 40 Gbd. In: 2013 IEEE Int. Top. Meet. Microw. Photonics. IEEE; 2013. pp. 119-122. DOI: 10.1109/MWP.2013.6724034

[39] Timoshenko S, Goodier J. Theory of Elasticity. New York: McGraw-Hill; 1970

[40] Kurashima T, Tateda M. Thermal effects on the Brillouin frequency shift in jacketed optical silica fibers. Applied Optics. 1990;**29**:2219-2222. DOI: 10.1364/AO.29.002219

[41] Horiguchi T, Kurashima T, Tateda M. Tensile strain dependence of Brillouin frequency shift in silica optical fibers. IEEE Photonics Technology Letters. 1989;**1**:107-108. DOI: 10.1109/68.34756

[42] Galindez-Jamioy CA, López-Higuera JM. Brillouin distributed fiber sensors: An overview and applications. Journal of Sensors. 2012;**2012**:1-17. DOI: 10.1155/2012/204121

[43] Zou W, He Z, Kishi M, Hotate K. Stimulated Brillouin scattering and its dependences on strain and temperature in a high-delta optical fiber with F-doped depressed inner cladding. Optics Letters. 2007;**32**:600. DOI: 10.1364/OL.32.000600

[44] Zou W, He Z, Hotate K. Investigation of strain- and temperature-dependences of Brillouin frequency shifts in GeO2-doped optical fibers. Journal of Lightwave Technology. 2008;**26**:1854-1861. DOI: 10.1109/JLT.2007.912052

[45] Bernini R, Minardo A, Zeni L. Reconstruction technique for stimulated Brillouin scattering distributed fiber-optic sensors. Optical Engineering. 2002;**41**:2186-2194. DOI: 10.1117/1.1497176

[46] Kurashima T, Horiguchi T, Izumita H, Furukawa S, Koyamada Y. Brillouin optical-fiber time domain reflectometry. IEICE Transactions on Communications. 1993;**E76-B**:382-390

[47] Song K-Y, He Z, Hotate K. Distributed strain measurement with Millimeter-order spatial resolution based on Brillouin optical correlation domain analysis and beat lock-in detection scheme. In: Opt. Fiber Sensors. Washington, D.C.: OSA; 2006. p. ThC2. DOI: 10.1364/OFS.2006.ThC2

[48] Song K-Y, Hotate K. Brillouin optical correlation domain analysis in linear configuration. IEEE Photonics Technology Letters. 2008;**20**:2150-2152. DOI: 10.1109/LPT.2008.2007744

[49] Garus D, Krebber K, Schliep F, Gogolla T. Distributed sensing technique based on Brillouin optical-fiber frequency-domain analysis. Optics Letters. 1996;**21**:1402-1404. DOI: 10.1364/OL.21.001402

[50] Zadok A, Zilka E, Eyal A, Thévenaz L, Tur M. Vector analysis of stimulated Brillouin scattering amplification in standard single-mode fibers. Optics Express. 2008;**16**:21692-21707. DOI: 10.1364/OE.16.021692

[51] Motil A, Bergman A, Tur M. State of the art of Brillouin fiber-optic distributed sensing. Optics and Laser Technology. 2016;**78**:81-103. DOI: 10.1016/j.optlastec.2015.09.013

[52] M a S, Thévenaz L. Modeling and evaluating the performance of Brillouin distributed optical fiber sensors. Optics Express. 2013;**21**:31347-31366. DOI: 10.1364/OE.21.031347

[53] Lopez-Gil A, Soto MA, Angulo-vinuesa X, Dominguez-Lopez A, Martin-Lopez S, Thévenaz L, et al. Evaluation of the accuracy of BOTDA systems based on the phase spectral response. Optics Express. 2016;**24**: 17200-17214. DOI: 10.1364/OE.24.017200

[54] Horiguchi T, Shimizu K, Kurashima T, Tateda M, Koyamada Y. Development of a distributed sensing technique using Brillouin scattering. Journal of Lightwave Technology. 1995; **13**:1296-1302. DOI: 10.1109/50.400684

[55] Lecoeuche V, Webb DJ, Pannell CN, D a J. Transient response in high-resolution Brillouin-based distributed sensing using probe pulses shorter than the acoustic relaxation time. Optics Letters. 2000;**25**:156. DOI: 10.1364/OL.25.000156

[56] Tai K, Hasegawa A, Tomita A. Observation of modulational instability in optical fibers. Physical Review Letters. 1986;**56**:135-138. DOI: 10.1103/PhysRevLett.56.135

[57] Alem M, Soto MA, Thévenaz L. Analytical model and experimental verification of the critical power for modulation instability in optical fibers. Optics Express. 2015;**23**:29514-29532. DOI: 10.1364/OE.23.029514

[58] Tai K, Tomita A, Jewell JL, Hasegawa A. Generation of subpicosecond solitonlike optical pulses at 0.3 THz repetition rate by induced modulational instability. Applied Physics Letters. 1986;**49**:236-238. DOI: 10.1063/1.97181

[59] Van Simaeys G, Emplit P, Haelterman M. Experimental demonstration of the Fermi-Pasta-Ulam recurrence in a modulationally unstable optical wave. Physical Review Letters. 2001;**87**:033902. DOI: 10.1103/PhysRevLett.87.033902

[60] Thévenaz L, Mafang SF, Lin J. Effect of pulse depletion in a Brillouin optical time-domain analysis system. Optics Express. 2013;**21**:14017-14035. DOI: 10.1364/OE.21.014017

[61] Ruiz-Lombera R, Urricelqui J, Sagues M, Mirapeix J, López-Higuera JM, Loayssa A. Overcoming nonlocal effects and Brillouin threshold limitations in Brillouin optical time-domain sensors. IEEE Photonics Journal. 2015;**7**. DOI: 10.1109/JPHOT.2015.2498543

[62] Soto MA, Alem M, Chen W, Thévenaz L. Mitigating modulation instability in Brillouin distributed fibre sensors. In: Jaroszewicz LR, editor. Proc. SPIE, Fifth Eur. Work. Opt. Fibre Sensors. Vol. 8794. Krakow, Poland; 2013. p. 87943J. DOI: 10.1117/12.2026296

[63] Dong Y, Bao X. High spatial resolution and long-distance BOTDA using differential Brillouin gain in a

dispersion shifted fiber. In: 20th Int. Conf. Opt. Fibre Sensors. Vol. 7503. 2009. p. 750384-4. DOI: 10.1117/12.848676

[64] Foaleng SM, Thévenaz L. Impact of Raman scattering and modulation instability on the performances of Brillouin sensors. Proceedings of SPIE The International Society for Optical Engineering. 2011;**7753**:77539V-775394V. DOI: 10.1117/12.885105

[65] Soto MA, Ricchiuti AL, Zhang L, Barrera D, Sales S, Thevenaz L. Time and frequency pump-probe multiplexing to enhance the signal response of Brillouin optical time-domain analyzers. Optics Express. 2014; **22**:28584-28595. DOI: 10.1364/OE.22.028584

[66] Voskoboinik A, Yilmaz OF, Willner AW, Tur M. Sweep-free distributed Brillouin time-domain analyzer (SF-BOTDA). Optics Express. 2011;**19**: B842-B847. DOI: 10.1364/OE.19.00B842

[67] Urricelqui J, Alem M, Sagues M, Thévenaz L, Loayssa A, Soto MA. Mitigation of modulation instability in Brillouin distributed fiber sensors by using orthogonal polarization pulses. In: Proc SPIE, 24th Int Conf Opt Fibre Sensors. Vol. 9634. 2015. p. 963433. DOI: 10.1117/12.2195292

[68] Minardo A, Bernini R, Zeni L, Thevenaz L, Briffod F. A reconstruction technique for long-range stimulated Brillouin scattering distributed fibre-optic sensors: Experimental results. Measurement Science and Technology. 2005;**16**:900-908. DOI: 10.1088/0957-0233/16/4/002

[69] Zornoza A, Minardo A, Bernini R, Loayssa A, Zeni L. Pulsing the probe wave to reduce nonlocal effects in brillouin optical time-domain analysis (BOTDA) sensors. IEEE Sensors Journal. 2011;**11**:1067-1068. DOI: 10.1109/JSEN.2010.2078805

[70] Dominguez-Lopez A, Angulo-Vinuesa X, Lopez-Gil A, Martin-Lopez S, Gonzalez-Herraez M. Non-local effects in dual-probe-sideband Brillouin optical time domain analysis. Optics Express. 2015;**23**:10341. DOI: 10.1364/OE.23.010341

[71] Iribas H, Urricelqui J, Mompó JJ, Mariñelarena J, Loayssa A. Non-local effects in Brillouin optical time-domain analysis sensors. Applied Sciences. 2017; 7:761. DOI: 10.3390/app7080761

[72] Bernini R, Minardo A, Zeni L. Long-range distributed Brillouin fiber sensors by use of an unbalanced double sideband probe. Optics Express. 2011; **19**:23845. DOI: 10.1364/OE.19.023845

[73] Mompó JJ, Urricelqui J, Loayssa A. Brillouin optical time-domain analysis sensor with pump pulse amplification. Optics Express. 2015;**24**:1340-1348. DOI: 10.1364/OE.24.012672

[74] Urricelqui J, Sagues M, Loayssa A. Synthesis of Brillouin frequency shift profiles to compensate non-local effects and Brillouin induced noise in BOTDA sensors. Optics Express. 2014;**22**:18195. DOI: 10.1364/OE.22.018195

[75] Iribas H, Mariñelarena J, Feng C, Urricelqui J, Schneider T, Loayssa A. Effects of pump pulse extinction ratio in Brillouin optical time-domain analysis sensors. Optics Express. 2017;**25**: 27896-27911. DOI: 10.1364/OE.25.027896

[76] Feng C, Iribas H, Marinelaerña J, Schneider T, Loayssa A. Detrimental effects in Brillouin distributed sensors caused by EDFA transient. In: Conf. Lasers Electro-Optics; San Jose, Califonia, United States. 2017. p. JTu5A.85. DOI: 10.1364/CLEO_AT.2017. JTu5A.85

[77] Zornoza A, Olier D, Sagues M, Loayssa A. Brillouin distributed sensor using RF shaping of pump pulses.

Measurement Science and Technology. 2010;**21**:094021. DOI: 10.1088/0957-0233/21/9/094021

[78] Li W, Bao X, Li Y, Chen L. Differential pulse-width pair BOTDA for high spatial resolution sensing. Optics Express. 2008;**16**:21616. DOI: 10.1364/OE.16.021616

[79] Motil A, Danon O, Peled Y, Tur M. High spatial resolution BOTDA using simultaneously launched gain and loss pump pulses. In: Fifth Eur Work Opt Fibre Sensors. 2013. p. 87943L. DOI: 10.1117/12.2026670

[80] Kishida K, Li C. Pulse pre-pump-BOTDA technology for new generation of distributed strain measuring system. Structural Health Monitoring and Intelligent Infrastructure. 2005;**1**: 471-477

[81] Voskoboinik A, Wang J, Shamee B, Nuccio SR, Zhang L, Chitgarha M, et al. SBS-based Fiber optical sensing using frequency-domain simultaneous tone interrogation. Journal of Lightwave Technology. 2011;**29**:1729-1735. DOI: 10.1109/JLT.2011.2145411

[82] Bernini R, Minardo A, Zeni L. Dynamic strain measurement in optical fibers by stimulated Brillouin scattering. Optics Letters. 2009;**34**:2613-2615. DOI: 10.1364/OL.34.002613

[83] Jin C, Guo N, Feng Y, Wang L, Liang H, Li J, et al. Scanning-free BOTDA based on ultra-fine digital optical frequency comb. Optics Express. 2015;**23**:5277-5284. DOI: 10.1364/OE.23.005277

www.ingramcontent.com/pod-product-compliance
Lightning Source LLC
Chambersburg PA
CBHW081234190326
41458CB00016B/5784